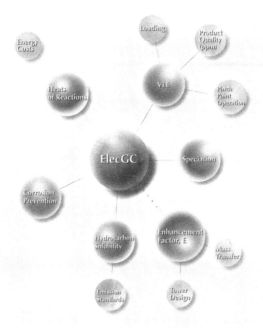

MOLECULAR THERMODYNAMICS
OF ELECTROLYTE SOLUTIONS

MOLECULAR THERMODYNAMICS
OF ELECTROLYTE SOLUTIONS

Lloyd L Lee

California State Polytechnic University, USA

 **World Scientific**

NEW JERSEY · LONDON · SINGAPORE · BEIJING · SHANGHAI · HONG KONG · TAIPEI · CHENNAI

Published by

World Scientific Publishing Co. Pte. Ltd.

5 Toh Tuck Link, Singapore 596224

USA office: 27 Warren Street, Suite 401-402, Hackensack, NJ 07601

UK office: 57 Shelton Street, Covent Garden, London WC2H 9HE

British Library Cataloguing-in-Publication Data
A catalogue record for this book is available from the British Library.

MOLECULAR THERMODYNAMICS OF ELECTROLYTE SOLUTIONS

ISBN-13 978-981-281-418-0
ISBN-10 981-281-418-3
ISBN-13 978-981-281-419-7 (pbk)
ISBN-10 981-281-419-1 (pbk)

Printed in Singapore.

Preface

Electrolytes and salt solutions are ubiquitous in chemical industry, biology, and nature. However, their thermodynamic properties and applications have not been adequately covered in the undergraduate curricula. Possibly, it was due to the theoretical level and difficulty in treating long-range Coulomb interactions; and partly it was due to the historical emphasis on hydrocarbon processing for oil-and-gas industry. Into the 21st century, the chemical industry becomes highly specialized and much diversified. Many new processes and products are developed that require the application of the electrolyte solutions and knowledge of their properties.

This book is written for the purposes of a learning primer on electrolyte solutions, connection to the modern molecular approaches in the field, and giving examples of some important industrial applications. There is in actuality a dearth of introductory textbooks on electrolyte solutions. The earlier books[*] by Harned and Owen and by Robinson and Stokes, valuable as they are, were written in the 1950s and 60s. They do not contain many of the modern materials. Other recent books are at an advanced level and restricted to specialized topics. There is indeed a need for a general introductory book to serve as the first course on electrolyte thermodynamics as well as a beginner's guide to it.

How do we harmonize the diametrically opposed demands of, on the one hand, an introduction and on the other, advanced molecular theories? We divide the book into three parts. The first five chapters are introductory, thus suitable for undergraduate education. They also provide practicing engineers who did not have training in this subject with a quick self study to "catch up". It can supplement an undergraduate thermodynamics course. Chapters 6 to 10 pave the way to molecular theories. We are heavy on applied statistical mechanics and light on

---

[*]H. S. Harned, B. B. Owen, *"The physical chemistry of electrolyte solutions"* (Reinhold Publishing, New York, 1950). R. A. Robinson, R. H. Stokes, *"Electrolyte solutions"* (Butterworths, London, 1959).

advancing theories. Chapters 12 and 13 are industrial applications: absorption refrigeration and natural gas processing; both utilizing electrolyte solutions. We give more description below.

We introduce the classical Debye-Hückel theory, not due to its accuracy, but for historical reasons. The contribution of Debye and Hückel is remarkable in their ability to explain the $(-\sqrt{I})$ behavior of the activity coefficients (I being the ionic strength). This behavior arises due to the long-range Coulomb interaction and cannot be accounted for by classical thermodynamic theories. It is now recognized as the *Debye-Hückel limiting law*. For higher concentrations, we choose the formulation due to Pitzer. Pitzer has correlated salt solution data for many common salts with a virial type equation. The equation is relative simple and can be programmed on a hand-held calculator (e.g. a TI-89). For simple systems, the results are quite accurate. It is hoped that after the five chapters, the reader will acquire adequate working skills with salt solutions and be able to apply equations to calculate accurately electrolyte solution properties with confidence.

To understand modern electrolyte theory, we must learn the molecular aspects of ionic solutions. The next five chapters (6 to 10) form an advanced part of the book and are more suitable for a graduate level course. It introduces the statistical mechanics of electrolyte solutions. Some elementary knowledge of physical chemistry, such as probability distribution, partition function, and statistical ensembles, is needed as prerequisites. Here we go beyond the Debye-Hückel theory and discuss the integral equation approaches that give more accurate activity coefficients for concentrated electrolyte solutions (greater than 0.001 molal). We also try to keep the mathematics on a moderate level, by using the mean spherical approach (MSA) as the staple method where analytical formulas are available for numerical calculations. Chapter 10 gives a general description of the Ornstein-Zernike integral equation for the molecular distributions. We describe the hypernetted-chain closure (HNC), the numerical solution techniques, and the origin and treatment of the bridge function that is essential in any closure equation. Once the integral equations are solved, the correlation functions can be used to obtain the thermodynamic properties of the electrolyte solution: such as the electrostatic internal energy, the osmotic coefficient, and the activity coefficient. Since the activity coefficients play a central role, just as in the conventional solution thermodynamics, we can decipher the behavior of electrolyte solutions from these properties. The reader should be able to connect comfortably with other modern treatments with a molecular bend after studying these chapters.

Many industrial processes employ mixed solvents: water, methanol, amines, ethylene glycol, ammonia, etc. in the presence of salt species, such as in inhibition of gas hydrates, dehydration, gas sweetening, azeotrope distillation, and refrigeration. Chapters 8 and 9 deal with the "salting-out" behavior for these solvents and the necessary conversion of thermodynamic scales: from the McMillan-Mayer scale to the Lewis-Randall scale. Our approach automatically satisfies the Gibbs-Duhem relation and thermodynamic consistency. We examine the Setchenov and Furter coefficients, which are rederived, improved, and put on a firm theoretical basis.

Chapter 11 is an introduction to the electric double layers that are at the basis of cellular interactions, biochemistry, and electrochemistry on electrodes. We start with the Poisson-Boltzmann equation and go into some details on the derivations to achieve a sound understanding. We then discuss the ζ-potential and the DLVO (Derjaguin-Landau-Verwey-Overbeek) theory that are much used in colloid and biochemistry. Finally we cite some recent developments in protein chemistry on using the Yukawa potentials to model the colloidal interactions. The molecular approach introduced in Chapter 10 finds its application here.

A unique feature of this text book is the inclusion of two chapters (12 and 13) on industrial applications: (i) the absorption refrigeration using electrolytes as working fluids, and (ii) the removal, from raw natural gas streams, of the acid gases (hydrogen sulfide and carbon dioxide) by aqueous amine solutions. There is urgent need in the HVAC (heating, ventilation, & air conditioning) industry in finding alternative working fluids other than the CFC's (chlorofluorocarbons) and HCFC's (hydrochlorofluorocarbons) that are ozone depleting and are subject to banning by 2030 according to an international accord. Ionic solutions, such as water-lithium bromide solutions, can be use in the absorption refrigeration cycle to achieve cooling. In natural gas processing, the acid gases must be neutralized and removed before being used as fuel. Aqueous amines, which contain ionic species, are used to "wash" and "sweeten" the raw gas, to remove the CO_2 and H_2S. This process also has implications in environmental engineering. We have included some software in the CD accompanying this book for calculations of the loading of amines and acid gas vapor pressures. It can be put on a laptop computer by an engineer and carried around in the field for fast answers. The programs also provide detailed outputs (such as speciation and activity coefficient information).

This text book is introductory. However it connects to modern approaches. It lays out the basic theories, but also gives industrial applications. It provides ready-to-use software[*] and computer programs that give immediate applications and alleviate the complicated programming. We hope that it will enhance the thermodynamics education and put the tools of accurate electrolyte modeling into the hands of practicing engineers.

The field of electrolyte solutions is vast and spanned many centuries. It is impossible to cite, let alone study, all the extant literature. Due to the introductory nature of this book, we mentioned only a bare minimum of previous works in the bibliography. The sin of omission is not to be interpreted as a lack of respect for all the brilliant and indispensable contributions to this field. I sincerely thank the careful review by Dr. Frank T.H. Chung who showed keen interest in its success, and many typing and editing help by my dear wife, Chi-Ming. Some of the programming has been carried out by my former colleagues, Kevin Gering, D. J. Ghonasgi, Lester Landis, Bill Li-Jun Lee, Frank Chung, and Rong-Song Wu. Many of them have pioneered in the developments of this branch of electrolyte solution studies. I am grateful for their contributions.

Pomona
California
Spring 2008

LLOYD L. LEE

---

[*]A Windows-interactive (GUI) software for acid gas treating with amine solutions is available for distribution at cost. Contact profllee@yahoo.com for ordering.

Contents

Attached Compact Disk
    ElecGC: Executable PC Program for Acid Gas Treating
    MSA: Executable PC Program for MSA Activity Coefficients

Chapter 1

Introduction

1.1. Prolegomena

Salts, that are neutral when pure, such as the table salt sodium chloride $NaCl$, can dissociate (i.e. "electrolyze") into ions Na^{+1} and Cl^{-1} when dissolved in a strong dielectric solvents (such as water)—i.e. forming an aqueous electrolyte solution.

$$NaCl \underset{in\ water}{\rightarrow} Na^+ + Cl^- \qquad (1.1.1)$$

Na^{+1} and Cl^{-1} are ions that have either a positive charge (called cations) or a negative charge (anions). Cl^{-1} is a chlorine atom with one added extra electron (electron carries one negative charge, at $-1.609E$-19 Coulomb) while Na^+ is the sodium atom which has lost an electron (thus with a positive charge: $+1.609E$-19 Coulomb). The valence is the number z of electrons lost $(+z)$ or gained $(-z)$. The valence of Na^{+1} is plus one (losing one electron), while that of Cl^{-1} is minus one (gaining one electron). For the salt Na_2SO_4, sodium ion Na^{+1} has valence again of plus 1, while SO_4^{-2} has valence of minus two. Therefore we have the valence types of salts based on the resulting valences of the ions. $NaCl$ is the type 1-1 electrolyte (namely, valence 1 for cation and valence -1 for anion), while Na_2SO_4 is type 1-2 electrolyte (valence 1 for cation and valence -2 for anion). We use the symbols z_+ or z_- to denote the valences of the cation and anion, respectively. The dissociation eq. (1.1.1) is a chemical reaction. All rules for chemical reactions apply equally to the case of electrolysis. For any salt CA (C = cation and A= anion), the dissociation chemical equilibrium is written as

$$CA = v_+ C^{z+} + v_- A^{z-} \qquad (1.1.2)$$

where v_+ and v_- are the stoichiometric coefficients for the cation and the anion, respectively.

In order for the salts to ionize, the solvent must have a high dieletric constant (to be defined in the following). At room temperature the relative dielectric constant D for water is about 80, and for diethylformamide 180. The higher the dielectric constant the easier it is for salts to dissociate. Salts can also dissociate as is (namely without solvents) at high temperatures, i.e., in the molten state at above $1000°C$ (the melting point depends on the salt). At room temperature, a dielectric solvent (or mixtures of many solvents) with high value of D is needed for dissociation. There are many salt types. We exhibit some examples below.

$$
\begin{aligned}
Na_2SO_4 &= 2Na^{+1} + SO_4^{-2} \\
CaSO_4 &= Ca^{+2} + SO_4^{-2} \\
AlCl_3 &= Al^{+3} + 3Cl^{-1} \\
Na_3PO_4 &= 3Na^{+1} + PO_4^{-3}
\end{aligned}
\tag{1.1.3}
$$

In the absence of an electrical field, the valences of ions z_+, z_- and the numbers of ions v_+, v_- combined will maintain the neutrality (zero net charge) of the whole solution.

$$
v_+z_+ + v_-z_- = 0
\tag{1.1.4}
$$

This is called the *electroneutrality principle*, which we shall use often later. For example, the table salt, NaCl (eq. (1.1.1)), has $z_+=+1$, $z_-=-1$, $v_+=1$, and $v_-=1$. The sum $v_+z_+ + v_-z_- = (1)(+1) + (1)(-1) = 0$. For Na_2SO_4, $z_+=+1$, $z_-=-2$, $v_+=2$, and $v_-=1$. The sum $v_+z_+ + v_-z_- = (2)(+1) + (1)(-2) = 0$. This electroneutrality applies to all valence types of salt solutions. For $MgSO_4$, a 2-2 electrolyte (valence 2 for cation Mg^{+2} and valence -2 for anion SO_4^{-2}): $z_+=+2$, $z_-=-2$, $v_+=1$, and $v_-=1$; for $AlCl_3$ a 3-1 electrolyte (valence 3 for cation Al^{+3} and valence -1 for anion): $z_+=+3$, $z_-=-1$, $v_+=1$, and $v_-=3$, and for Na_3PO_4, a 1-3 electrolyte (valence 1 for cation and valence -3 for anion PO_4^{-3}): $z_+=+1$, $z_-=-3$, $v_+=3$, and $v_-=1$. All types obey the electroneutrality rule.

Electrolyte solutions are mixtures consisting of dissociated ions in a solvent. Thus at equilibrium, the solution thermodynamic principles apply. Namely, the free energy is at a minimum, and the chemical potentials of species balance out.

$$\mu_{CA} = v_+\mu_+ + v_-\mu_- \tag{1.1.5}$$

where μ_+ and μ_- are the chemical potentials of the cation and anion respectively. The chemical potentials in the forward reaction are balanced by the chemical potentials in the backward reaction. A second remark is that at the equilibrium (1.1.5), the dissociation of *CA* is not always complete. If the dissociation is almost complete, such as *NaCl* in water, the salt is called a *strong electrolyte*; if the extent of dissociation is limited, the salt is called a *weak electrolyte*, such as $Ca(C_2H_5COO)_2$. We can tell the dissociation strength by examining the dissociation constant (the equilibrium constant).

Since the ions are charged species, so the electrostatic principles apply. Let us first explain the units. The charge of an electron is negative and is equal to -1.60218×10^{-19} Coulomb, or in electrostatic units, -4.803×10^{-10} esu. We use the symbol e to represent the *absolute* value of the charge of one electron (i.e., $e = +1.60218 \times 10^{-19}$ Coulomb). The permittivity of vacuum is $\varepsilon_0 = 111.265 \times 10^{-12}$ in units of $(Coulomb)^2/(Nm^2)$. It is important to recognize that in electrostatics there are alternative units and definitions. The permittivity value above is based on the following form of *Coulomb's law*

$$\vec{F}_{12} = \frac{1}{\varepsilon_m}\frac{q_1 q_2}{r^2}\vec{e}_{12} \tag{1.1.6}$$

where F_{12} is the force between two bodies 1 and 2 of charges q_1 and q_2 Coulombs, separated by a distance r_{12}, while e_{12} is the unit vector from charge 1 to charge 2. ε_m is called the *permittivity*, i.e., the proportionality constant. In different media (air, water, vacuum, or oil) where the two charges are immersed, the permittivity will have different values. The force can be expressed in Newton, charges in Coulomb, and distance in meter. However, alternative definition of the permittivity in Coulomb's law has been used (e.g. adding a factor 4π, to account for the spherical solid angle).

$$\vec{F}_{12} = \frac{1}{4\pi\varepsilon_m^+}\frac{q_1 q_2}{r^2}\vec{e}_{12} \qquad (1.1.7)$$

This definition gives a permittivity (with superscript +) of vacuum $\varepsilon_0^+ =$ $111.265 \times 10^{-12}/4\pi = 8.85419 \times 10^{-12}$ (Coulomb^2)/(Nm^2). It is always a good practice to verify which of the definitions of the permittivity is used at the beginning of any calculation.

In addition to the permittivity, a relative dielectric constant D is also used to characterize the dielectric medium. This relative dielectric constant is the ratio of the permittivity ε_m of the medium at hand to the permittivity ε_0 in vacuum

$$D \equiv \frac{\varepsilon_m}{\varepsilon_0} \qquad (1.1.8)$$

Thus D is a ratio and is dimensionless. For vacuum, $D = 1$. The permittivity of air is almost the same as vacuum, $D_{air} \sim 1$. Given the value of the relative dielectric constant D, we can always recover the actual permittivity by $\varepsilon_m = (D\,\varepsilon_0)$. One can find the relative dielectric constants in many handbooks (such as the CRC Handbook on physical chemistry.[107] See also Appendix III). The relative dielectric constant is a function of temperature. For example, water has $D = 80.176$ (at 20°C), 78.358 (at 25°C), and 76.581 (at 30°C). On the other hand, for methanol $D = 33$ (at 25° C), ethanol $D = 24.3$ (at 25° C) and ammonia $D = 17$ (at 20°C). To obtain the permittivity of water at 20°C, $\varepsilon_{water} = D_{water}\,\varepsilon_0 = 80.176 \times 111.265\text{E-}12 = 8920.78\text{E-}12$ (Coulomb)^2/(Nm^2).

Older literature uses the terms permittivity or dielectric constant, interchangeably. Thus when "dielectric constant" was mentioned, it could mean the *relative dielectric constant, D,* or the permittivity ε_m. Furthermore, the word dielectric constant was also used either for the *permittivity* ε_m^+ or ε_m, one with, the other without the factor 4π. We need to exercise caution when reading the literature.

1.2. Concentration Units

When salts are dissolved in water, its concentration is expressed (in the electrolyte solution literature) by at least three different unit systems: (i) the practical units (*molality*), (ii) the rational units (*mole fractions*), and (iii) the molar units (*molarity*). Mole fractions are easy to understand.

However, for the convenience in experimental work in physical chemistry, the practical units have been devised.

Practical Unit-- Molality, M: In the laboratory, one can easily measure 1000 g (1 kg) of pure water, then measure 58.4428g of sodium chloride (molecular weights: *Na* 22.9898+ *Cl* 35.453) After mixing the resulting a salt solution is called *1 molal* (or 1.0M). Thus molality *M* scale is defined as

$$M = \frac{n_s}{Wt\ kg\ H_2O} = \frac{no.\ of\ gmoles\ of\ salt}{No.\ of\ kg\ H_2O} \tag{1.2.1}$$

Molar Units-- Molarity, c: On the other hand, if one puts emphasis on the volume produced, one would weigh 58.4428g of table salt in a graduate, and pour in water until the total volume (salt+water) reaches 1000 cc. (For a graduate with smaller volume, one can do it proportionally: e.g. 5.84428g salt for a 100 cc graduate). This mixture is at *1 molar*. Note that the final volume one liter is for a mixture of both salt and water (not just for pure water at 1000 g). Thus

$$c = \frac{n_s}{liter} = \frac{no.\ of\ gmoles\ of\ salt}{liter\ solution\ (water + salt)} \tag{1.2.2}$$

For low concentrations of salt, one liter of solution and 1000 g of pure water are similar in weight, thus $M \approx c$. But their difference becomes large at high salt concentrations. The conversion between the two units is given by:

Basis = 1000g of pure water

$$c = M\left(\frac{1000 d_m}{n_s W_s + 1000}\right) \tag{1.2.3}$$

and

$$M = c\left(\frac{n_s W_s + 1000}{1000 d_m}\right) \tag{1.2.4}$$

where d_m is the density (in kg/liter) of the salt solution at the given molarity

$d_m = (Wt_{solu}, total\ weight\ of\ solution,\ kg)/(V_{solu},\ volume\ of\ solution,\ liter)$

(Note: total weight of the solution $Wt_{solu} = (n_s\ W_s + 1000)g\ /1000)$, in kg)

$$d_m = \frac{n_s W_s + 1000}{1000\ V_{solu}\ (liter)} \tag{1.2.5}$$

where n_s = number of gmoles of the salt, and W_s = molecular weight of the salt (in g/gmol, e.g. W_s =58.4428g /gmol for *NaCl*), and V_{solu} = volume (in liter) of the solution (i.e., mixture). In dilute salt solutions, n_s ~0, thus d_m ~ d_0 = the density of pure water which is around 1 kg/liter. Thus $c \approx M$. In case the solvent is not water (for example, ethanol), we apply the same procedure, we shall have 1000 g (1 kg) of ethanol instead of water. In many electrolyte solution studied, water is the only solvent. So we shall implicitly assume that the solution is aqueous, unless otherwise specified.

Rational units: mole fractions, x_i:

The mole fraction is defined as usual. There are two pictures (or scales) for electrolyte solutions: the McMillan-Mayer (MM) scale (solvent-implicit scale) and the Lewis-Randall (LR) scale (solvent-explicit scale). In the MM scale, the solvent (water) molecules are absent ("removed"). In the LR scale all molecules are present, including the solvent water. (Details to be explained in Chapter 4).

McMillan-Mayer scale (solvent-implicit)

The mole fractions are based on the ions (cations and anions) present without participation of the water molecules. The number of moles of ions are n_i (i= Na$^+$, Cl$^-$, Mg^{+2}, SO$_4^{-2}$, etc.)

$$x_i = \frac{n_i}{\displaystyle\sum_{j=all\ ions} n_j} \tag{1.2.6}$$

Lewis-Randall scale (solvent-explicit)

In this scale, the solvent molecules are restored with moles $= n_0$. The mole fractions x_α' are defined as

$$x_\alpha' = \frac{n_\alpha}{n_0 + \sum\limits_{j=allions} n_j} \tag{1.2.7}$$

α is one of the ion species and can be the solvent species. The conversion from molality and molarity to mole fraction is simple.

[Example 1.1]: Given 125g of *NaCl* salt, dissolve it in 2 kg of water. The density of the solution (mixture) is 1.072 kg/liter at 25°C. Find the salt concentration in molality, molarity, and mole fraction.
Answer: First we find n_s the number of gmoles of salt,

$$n_s = 125 \ g/\ (58.4428g/gmole) = 2.139 \ gmoles$$

Thus the molality M

$$M = n_s\ /\ 2 \ kg \ water = 2.139/2 = 1.0695 \ molal.$$

And from (1.2.3)

$$c = M.(\ (1000 \ d_m)/(\ n_s \ W_s + 1000)\ ) =$$
$$= 1.0695 \ (1000 \ x \ 1.072)/(\ 1.0695 x 58.4428 + 1000) = 1.079 \ molar.$$

In the MM picture

$$x_{Na+} = 2.139/\ (2.139 + 2.319) = 0.50$$

So is

$$x_{Cl-} = 2.139/\ (2.139 + 2.319) = 0.50$$

In the LR picture

$$x'_{Na+} = 2.139/\ (2.139 + 2.319 + 2000/18) = 0.020$$
$$x'_{Cl-} = 2.139/\ (2.139 + 2.319 + 2000/18) = 0.020$$
$$x'_{water} = (2000/18)/\ (2.139 + 2.319 + 2000/18) = 0.96$$

At higher salt concentration, say 585g of *NaCl*, the molality would be 5.00 molal.   The density of solution is 1.166 g/cc at 25°C (from handbook[66]). The molarity would be

$$5((1000 \times 1.166)/(5 \times 58.4428+1000)) = 4.51 \ molar.$$

This value is very different from the molality of 5M (by −10%).

□

[Example 1.2]   The argon-argon interaction potential u_{AA} can be represented by dispersion force as in the Lannard-Jones potential.   The size parameter σ =3.405 Å, and the energy parameter ε =0.1654E-20 J. Compared to the Coulomb potential for *NaCl* solution in water, what is the ratio of the two energies? (The temperature is at 20°C).

Answer: The minimum energy −ε in the argon potential is at r_{min}=3.822 Å. We use this value in calculating the ion-ion interaction from the Coulomb potential

$$u_{12} \ = \ \frac{1}{\varepsilon_m} \frac{q_1 q_2}{r} \ = \ \frac{1}{(80.176)(111.2E-12)} \frac{(1.602E-19)^2}{(3.822E-10)} = (0.753E-20)J$$

Note that the relative dielectric constant *D* of water is 80.176 at 20°C, and the permittivity is 111.265×10^{-12}. Thus the ratio is

$$u_{12}/u_{AA} \ = \ (0.753E-20)/(0.1654E-20) = 4.6$$

The electrostatic energy at the same distance is 4.6 times stronger than the Lennard-Jones energy. In the case of a 2-2 electrolyte solution (e.g., $MgSO_4$), the valences are +2 and -2, ($q_1 = 2e$, and $q_2= -2e$), the energy ratio is 4.6 ×4= 18.4, eighteen times stronger. When these ions are in air, the relative dielectric constant *D* of air is ~1, the forces are 80 times stronger: 80.176×4.6 = 369 times for 1-1 electrolytes, and 80.176 × 18.4= 1475 times for 2-2 electrolytes. This shows that the electrostatic forces are very strong, hundred and thousand times stronger than the dispersion forces. This also explains the statement made earlier that we need strong dielectric solvents to enable dissociation of ions.   Only when the permittivity is large (D ~70 → 100), the interatomic force between the cation and the anion is weakened. This reduction of the attraction force enables the salt to dissociate into ions. Water (*D*=80) is a good solvent, thus many salt species can ionize in it. At room temperature T= 293.15K, the thermal fluctuation energy is proportional to ~kT:

$$kT = (1.38054 \text{ E-}23 \text{ J/K}) (293.15 \text{ K}) = 0.4 \text{ E-}20 \text{ Joule}$$

The thermal agitations are of the same order of the Coulomb attraction $(0.75 \text{ E-}20 \text{ J})$. Thus salt can ionize in water. Ethanol $(D = 24.3)$ is a weaker dielectric. Salts do not dissociate completely in ethanol. In air $(D = 1)$, the electrostatic forces are so strong, that at room temperature, salts rarely separate into ions.

□

Exercises:

1.1. Show, by your own reasoning, the conversion formulas from molarity c to molality M. (a) Take a basis of 1000 g of water. (b) Take at basis 1 liter of solution.

1.2. Given the density of aqueous solution of KCl = 1.1575 kg/liter at 3.9618 M $(10°C)$, find the molarity, c.

1.3. Find the force F in Newton between two charged bodies with $q_1 = 2$ Coulomb, and $q_2 = 3.5$ Coulomb in a dielectric medium of relative dielectric constant $D = 2.8$. The distance between the two bodies is 0.75 meter.

1.4. What ions are formed when you dissolve the salts NH_4I, Na_2S, $NaNO_4$, and LiBr in water? What are the valence types of these salts?

1.5. Find the molarity and mole fractions of salt $CuSO_4$ in water at $15°C$. (a)1.005M (density = 1.1573 g/cc). (b) 1.265M (density = 1.1965 g/cc). Find both x_i and x_i'.

Chapter 2

Solution Thermodynamics of Electrolytes

In solution thermodynamics, the quantity *par excellence* is the activity coefficient, which gives the non-ideal behavior of the solute species. Since electrolyte solutions obey the same thermodynamics, we shall give a brief review of solution thermodynamics.

There are two levels of formulas: the *fundamental* ones and the *practical* ones. We start with the fundamental formulas, then specialize to the conventional ones. The chemical potential μ_i of component i in a mixture is the partial molar Gibbs free energy

$$\mu_i = \left[\frac{\partial G}{\partial n_i} \right]_{T,P,n_{j \neq i}} \tag{2.1}$$

The fugacity f_i can then be defined as

$$\mu_i \equiv \mu_i^R + kT \ln f_i \tag{2.2}$$

where μ_i^R is a temperature-dependent ideal-gas reference chemical potential. The boundary condition on (2.2) is that as the system pressure $P_{sys} \rightarrow 0$ (low pressures), the fugacity f_i should approach the partial pressure, $f_i \rightarrow p_i = x_i P_{sys}$. From statistical mechanics of the ideal gas (*idg*), we know that

$$\mu_i^{idg} \equiv kT \ln \frac{\Lambda_i^3}{kT} + kT \ln p_i \tag{2.3}$$

where $p_i = x_i P_{sys}$ is the partial pressure. Thus the reference chemical potential is identified as

$$\mu_i^R = kT \ln \frac{\Lambda_i^3}{kT}\Big|_i \qquad (2.4)$$

Note that this form is limited to monatomic species with only the translational kinetic energy. For structured molecules, the other degrees of freedom (i.e. rotational and vibrational kinetic energies) should be included. We shall add a term q_{rv} to represent both the rotational and the vibrational contributions. Based on the input from statistical mechanics[59], the fugacity formula can now be written as

$$\mu_i \equiv kT \ln \frac{q_{rv}\Lambda_i^3}{kT} + kT \ln f_i \qquad (2.5)$$

A remark on our derivations is in order: first from the Gibbs free energy we have defined the chemical potential, and from the chemical potential, we have defined the fugacity. Now we want to define two more "convenient" quantities, the activity, a_i, and the activity coefficient, γ_i. The activity is defined at temperature T and pressure P as the ratio of the fugacity of component i in the mixture of concentration **x**, (**x** being the vector of mole fractions), to the fugacity of i at a reference state (*Ref*).

$$a_i \equiv \frac{f_i}{f_i^{\mathrm{Re}f}} \qquad (2.6)$$

Since we have the freedom (or expediency) of choosing the reference fugacity, f_i^{Ref}, the value of the activity can change according to the chosen reference, however, the fugacity f_i, being a fundamental quantity, is well-defined, not subject to change with reference states. In literature there are at least two choices of the reference states: the symmetric and the asymmetric references. The symmetric reference is the *pure (p or 0)* component state (when $x_i = 1$), the asymmetric reference is the *infinite dilution (d or ∞)* state (when $x_i = 0$).

$$a_i^p \equiv \frac{f_i(x,T,P,liq.state)}{f_i^0(x_i=1,T,P,liq.state)} \qquad (2.7)$$

$$a_i^d \equiv \frac{f_i(x,T,P,liq.state)}{f_i^\infty(x_i=0,T,P,liq.state)} \tag{2.8}$$

The superscripts p denotes pure i, and d denotes infinite dilution of i. The word *liq.state* refers to the liquid state. Since in general, $f^0 \neq f^\infty$, thus the two activities a^p and a^d do not have the same value (all the while the fugacity f_i has a unique value and is physically well-defined). In thermodynamics, we also use an "activity coefficient", γ_i, defined as

$$\gamma_i \equiv \frac{a_i}{x_i} \tag{2.9}$$

Since there are two definitions for activity, there are also two corresponding values for the activity coefficient:

$$\gamma_i^p \equiv \frac{f_i(x,T,P,liq.state)}{x_i f_i^0(x_i=1,T,P,liq.state)} \tag{2.10}$$

$$\gamma_i^d \equiv \frac{f_i(x,T,P,liq.state)}{x_i f_i^\infty(x_i=0,T,P,liq.state)} \tag{2.11}$$

According to Henry's law, $f_i^\infty = K_i$, the Henry's law constant. Thus the two activity coefficients are related by

$$\gamma_i^p f_i^0(x_i=1,T,P,liq.state) = \gamma_i^d K_i(T,P,liq.state) \tag{2.12}$$

From these developments, the chemical potentials are expressed in either convention as:

$$\frac{\mu_i}{RT} \equiv \ln\frac{q_{rv}\Lambda_i^3}{kT} + \ln x_i + \ln f_i^0 + \ln\gamma_i^p \tag{2.13}$$

or

$$\frac{\mu_i}{RT} \equiv \ln\frac{q_{rv}\Lambda_i^3}{kT} + \ln x_i + \ln f_i^\infty + \ln\gamma_i^d = \ln\frac{q_{rv}\Lambda_i^3}{kT} + \ln x_i + \ln K_i + \ln\gamma_i^d \tag{2.14}$$

We note emphatically that the chemical potentials and fugacities are "*fundamental*" quantities, not dependent on the reference state chosen; while the activities and activities are "*derived*" (or convenient) quantities. The values of the latter depend on the reference fugacity f_i^{Ref}. In electrolyte solutions, as a rule, the infinite dilution convention is adopted. Namely the activity coefficient γ_i of ion i (i = + or –) is defined to be unity at infinite dilution $x_i = 0$. (One does NOT have a pure ion state under ordinary conditions). In the future, we shall drop the superscript d (infinite dilution) on the activity coefficients γ_i^d for ions. The activity coefficients defined thusly will have value = 1, when the salt is infinitely dilute $x_i = 0$ in the solution.

For example in aqueous *NaCl* solution, there are four species present in the solution: the salt: *NaCl*, the dissociated ions Na^+, and Cl^-. and water. Their activity coefficients are γ_{NaCl}, γ_+, γ_- , and γ_{water} respectively.. Since experiments do not measure single ion activity, say γ_+ only (though recent discussions have focused on single ion activity[111]), the measured value is often a mean value of the ions, thus we define the *mean activity coefficient* (MAC) $\gamma_\pm$ as

$$\ln \gamma_\pm \;\equiv\; \frac{v_+ \ln \gamma_+ + v_- \ln \gamma_-}{v_+ + v_-} \tag{2.15}$$

We have given the activities above in terms of mole fractions. Since there are other concentration units, the activity coefficients can also be defined in units other than mole fractions.. Eq.(2.9) can now be formulated on the molality (the *practical units*), or molarity bases

$$\gamma_i^M \;\equiv\; \frac{a_i}{M_i} \tag{2.16}$$

$$\gamma_i^c \;\equiv\; \frac{a_i}{c_i} \tag{2.17}$$

The superscripts indicate that the concentration unit is in molality (M), or molarity (c). (Recall that we refer all reference fugacities to the infinite dilution of the salt species). Thus the chemical potentials are expressed in either M or c scale as

$$\frac{\mu_i}{kT} \;\equiv\; \ln \frac{q_{rv} \Lambda_i^3}{kT} + \ln M_i + \ln f_i^{\infty,M} + \ln \gamma_i^M \tag{2.18}$$

$$\frac{\mu_i}{kT} \equiv \ln\frac{q_{rv}\Lambda_i^3}{kT} + \ln c_i + \ln f_i^{\infty,c} + \ln \gamma_i^c \qquad (2.19)$$

Similarly, the mean activity coefficient can be defined for $\gamma_\pm^M$ and $\gamma_\pm^c$ as in (2.15). Now we have three concentration scales: M, c, and x. One can relate one to the other via unit conversion. These complications arise from the historical choices of concentration units. We show some results here and leave the rest to an exercise. As we know that at infinite dilution all activity coefficients tend to unity (by definition). Since the chemical potentials are the same, no matter what units are used,

$$\begin{aligned}
\frac{\mu_i^\infty}{kT} &= \lim_{xi\to 0}\ln\frac{q_{rv}\Lambda_i^3}{kT} + \ln x_i + \ln f_i^\infty = \\
&= \lim_{mi\to 0}\ln\frac{q_{rv}\Lambda_i^3}{kT} + \ln M_i + \ln f_i^{\infty,M} = \\
&= \lim_{ci\to 0}\ln\frac{q_{rv}\Lambda_i^3}{kT} + \ln c_i + \ln f_i^{\infty,c}
\end{aligned} \qquad (2.20)$$

Thus we have

$$\ln f_i^\infty = \ln f_i^{\infty,m} + \ln\frac{1000}{W_{H2O}} = \ln f_i^{\infty,c} + \ln\frac{1000 d_0}{W_{H2O}} \qquad (2.21)$$

Equation (2.21) converts one infinite dilution fugacity to the other. The fugacities are related consequently to one another as

$$\ln f_i = \ln x_i + \ln f_i^\infty + \ln \gamma_i = \ln M_i + \ln f_i^{\infty,m} + \ln \gamma_i^m = \ln c_i + \ln f_i^{\infty,c} + \ln \gamma_i^c \qquad (2.22)$$

By (2.22), we have the relations among the activity coefficients

$$\ln \gamma_i = \ln \gamma_i^m + \ln\frac{W_{H2O}M_i}{1000 x_i} = \ln \gamma_i^c + +\ln\frac{W_{H2O}c_i}{1000 d_0 x_i} \qquad (2.23)$$

If we express the activity coefficients in terms of mean quantities, $x_\pm$, $M_\pm$ and $c_\pm$ (as in eqs.(2.15 & 23)) becomes

$$\ln \gamma_\pm = \ln \gamma_\pm^m + \ln\frac{W_{H2O}M_\pm}{1000 x_\pm} = \ln \gamma_\pm^c + +\ln\frac{W_{H2O}c_\pm}{1000 d_0 x_\pm} \qquad (2.24)$$

where

$$\ln x_{\pm} \equiv \frac{v_{+} \ln x_{+} + v_{-} \ln x_{-}}{v_{+} + v_{-}},$$

$$\ln m_{\pm} \equiv \frac{v_{+} \ln m_{+} + v_{-} \ln m_{-}}{v_{+} + v_{-}}, \qquad (2.25)$$

$$\ln c_{\pm} \equiv \frac{v_{+} \ln c_{+} + v_{-} \ln c_{-}}{v_{+} + v_{-}}$$

Exercises:

2.1.   Find the activity coefficients[65] of aqueous $CuSO_4$ in molar and mole fraction units, given the molal activity coefficient $\gamma_{\pm}{}^{m} = 0.103$ at M=0.2M (25°C). The density of the solution is 1.0394 g/cc.

2.2.   Find the activity coefficients[65] of aqueous $(NH_4)_2SO_4$ in molar and mole fraction units, given the molal activity coefficient $\gamma_{\pm}{}^{m} = 0.116$ at M=4.0M (25°C). The density of the solution is 1.1959 g/cc.

2.3.   In acid gas treating, aqueous amine is used (amine = methyldiethanolamine (MDEA) + CO2 +water). The activity coefficients have been calculated as

$\gamma_{ion}{}^{m}$ for $(MDEA \cdot H^{+}, H_3O^{+}, HCO3^{-}, CO3^{=})$ = (1.485   1.0098   1.749   3.771), respectively .

Find the mean activity coefficient $\gamma_{\pm}{}^{m}$.

The reactions are

(MDEA)
$$[HO(CH_2)_2N]_2N\text{-}CH_3 + H_3O^{+} \longleftrightarrow [HO(CH_2)_2N]_2NH^{+}\text{-}CH_3 + H_2O$$

(CO$_2$ in water)

$$CO_2 + 2 H_2O \longleftrightarrow HCO_3^{-} + H_3O^{+}$$

$$HCO_3^{-} + H_2O \longleftrightarrow CO_3^{=} + H_3O^{+}$$

Chapter 3

Basic Electrostatics

In order to understand the interactions of charged species, some knowledge of the basic laws of electrostatics is required. We introduce below elementary principles of electrostatics: Coulomb's law, Poisson's law and Gauss' law. Reader interested in more details should consult a book on electrostatics.[21] The foundation for all three laws rests on the first law: Coulomb's law, which we have alluded to in eq.(1.1.6). The other two laws can be *derived* from (1.1.6).

3.1. Coulomb's Law

The force acting among two charges q_1 and q_2, separated by a distance r is given by

$$\vec{F}_{12} = \frac{1}{\varepsilon_m} \frac{q_1 q_2}{r^2} \vec{e}_{12} \qquad (3.1.1)$$

where ε_m is the *permittivity of the medium*. The interaction energy, by the rules of physics, is the integral of the force over the distance (from infinity to *r*). Thus the Coulomb electrostatic energy $u_\pm(r)$ is

$$u_\pm(r) = \frac{1}{\varepsilon_m} \frac{q_1 q_2}{r} \qquad (3.1.2)$$

In general, if other ions are present, we measure an *average* electrostatic energy $\phi(r)$.

3.2. Poisson's Law

The *Laplacian* of the electrostatic potential ϕ at r is proportional to the charge density at r

$$\nabla^2 \phi(r) = -\frac{4\pi}{\varepsilon_m} \rho_e(r) \qquad (3.2.1)$$

where $\rho_e(r)$ is the charge density (Coulomb per volume) at the distance r.

3.3. Gauss' Law

The electrical field $\underline{E_1}$ (Newton/Coulomb) on the surface of a body enclosing a distribution of electrical charges inside the body is given by

$$\iint_{cs} \vec{E_1} \bullet d\vec{S} = \sum_{i=1}^{k} \frac{4\pi q_i}{\varepsilon_m} \qquad (3.3.1)$$

where the double integrals are over the control surface CS and $\underline{E_1}$ is the electric field, defined as the negative gradient of the electrostatic potential per charge

$$\vec{E_1}(r) \equiv -\frac{\nabla \phi(r)}{q_1} \equiv -\nabla \phi_1(r) = \vec{F}_{12} / q_1 \qquad (3.3.2)$$

where we have defined $\nabla \phi_1$ (Newton/Coulomb) as the gradient of ϕ per charge q_1. Let us elaborate on the electric field. In electrostatics, the electric field at position r is the force experienced by a unit charge positioned at r. The test charge used can be one coulomb or one electron (depending on the units).

3.4. Relations Among the Three Laws of Electrostatics

Now we can relate all three laws of electrostatics via Green's theorems in calculus. We use Coulomb's law as the starting point.
[*Proof*]:
        Let us examine the geometry of two charges distributed in space as shown in Figure 3.1. We assume the control surface CS, without any loss of generality, to be spherical. The body (control volume, CV) of the

sphere contains an amount of charge q_2 inside. (We situate q_2 at the center, again for simplicity). Then if we put a unit test charge at a distance r from the center on the CS, the force on this test charge will be, according to Coulomb's law

$$\vec{E}_1 = \frac{\vec{F}_{12}}{q_1} = \frac{1}{\varepsilon_m} \frac{q_2}{r^2} \vec{n} \tag{3.4.1}$$

The direction of the field will coincide with the outward normal $\underline{n}$ of the surface (because of the spherical symmetry). The differential surface element is $d\underline{S} = \underline{n}.(rd\theta)(rsin(\theta))\ (d\phi)$ in spherical coordinates. The surface integral in (3.4) can be written in spherical coordinates as (since the dot product $\underline{n}.\underline{n} = 1$)

$$\iint_{cs} \vec{E}_1 \bullet d\vec{S} = \int rd\theta \int r\sin(\theta)d\phi \frac{q_2}{\varepsilon_m r^2} \vec{n} \bullet \vec{n} = \frac{4\pi\ q_2}{\varepsilon_m} \tag{3.4.2}$$

where the angular integrations gave a factor of 4π. Similarly, if we had multiple charges (k charges of q_2, q_3, ... etc.) inside the body (CV), we would have a summation of charges, Σq_i, not just q_2.

$$\iint_{cs} \vec{E}_1 \bullet d\vec{S} = \sum_{i=2}^{k} \frac{4\pi\ q_i}{\varepsilon_m} \tag{3.4.3}$$

This completes the proof of Gauss' law.

Let us define an electric charge density $\rho_e(r)$ (charge per volume, or Coulomb per cubic meter) as

$$\int\int\int_{CV} d\vec{r}\ \rho_e(r) \equiv \sum_{i=2}^{k} q_i \tag{3.4.4}$$

where we have replaced the sum of charges, Σq_i, by the volume integral of the electric charge density $\rho_e(r)$. Eq.(3.4.4) is a definition (or generalization) of the sum of charges Σq_i in terms of $\rho_e(r)$.

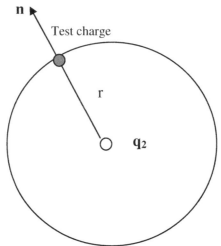

*Figure 3.1 A spherical control volume (CV) with an embedded charge q_2 at the center. A unit test charge is put on the surface at a radial distance r from q_2. **n** is the unit outward normal of the control surface (CS).*

To obtain Poisson's equation, we simply apply *Green's theorem* on the surface. The surface integral of the dot product of the vector $\mathbf{E_1}$ with the surface normal $\mathbf{n}$ (of d$\mathbf{S}$) should be equal to the volume integral of the gradient of $\mathbf{E_1}$. Using eqs.(3.3.2, 3.4.3 & 3.4.4), we have

$$\iint_{cs} \vec{E}_1 \bullet d\vec{S} = \iiint_{CV} d\vec{r}\, \frac{-\nabla \bullet \nabla \phi}{q_1} =$$

$$= -\iiint_{CV} d\vec{r}\, \nabla^2 \phi_1 = \sum_{i=2}^{k} \frac{4\pi q_i}{\varepsilon_m} = \iiint_{CV} d\vec{r}\, \frac{4\pi \rho_e(r)}{\varepsilon_m} \tag{3.4.5}$$

Since the control volume CV is arbitrary, and the equality hold for all surfaces considered, the integrands of both sides must be equal. This proves the Poisson equation (3.3).

$$\nabla^2 \phi_1(r) = -\frac{4\pi}{\varepsilon_m} \rho_e(r) \tag{3.4.6}$$

This equation is the basis of the Debye-Hückel equation of electrolyte solutions to be described next.

□

Exercises:

3.1. Given two bodies with charges $q_1 = 2.3$ Coulomb, $q_2 = -3.2$ Coulomb. They are placed at a distance 0.2 meter apart in air. What is the force acting between them?

3.2. Use Poisson's or Gauss equation, what is the electric field strength $\underline{E}_1$ (Newton/Coulomb) in air at a distance 1.5 *meters* from a charged body with 12 Coulomb?

3.3. Calculate the electric field strength $\underline{E}_1$ (Newton/Coulomb) at the surface of an ellipsoid with major/minor axes of *0.3m x 0.4m x 0.4m*, enclosing a charge q_2 at the origin with 26.5 Coulomb. The medium is air.

Chapter 4

The Debye-Hückel Theory

The activity coefficients of electrolyte solutions show a peculiar $(-\sqrt{I})$ (minus square root of I, I is the ionic strength) dependence at small salt concentrations. This dependence is rarely seen in the activity coefficients of neutral species. The origin of this negative square root dependence was first solved in the work of Debye and Hückel.[23] (1923). They made several ingenious simplifications based on the electrostatics and Boltzmann distribution. The result is the *Debye-Hückel limiting law* that firmly establishes this $(-\sqrt{I})$ dependence.

The Debye-Hückel theory is based on several suppositions: (i) they recognized that in aqueous electrolyte solutions there are cations, anions and water molecules existing side by side. However, water is a species with hydrogen-bonding energy. Furthermore the water molecules will hydrate the cations and anions in various ways. It was very difficult to account for the water structure precisely. Since the major physical effect of water in the electrolyte solution is to provide a "dielectric" buffer—to reduce the forces of Coulomb interaction among the ions, Debye and Hückel conceptually "removed" the water molecules from the mixture but retained the dielectric effect of water. The resulting physical space is called a "dielectric continuum". This way of looking at the electrolyte solution is called the *McMillan-Mayer picture*, where water molecules are stripped away. In reality, it corresponds to an *osmotic system*. (ii) The ions and counterions (ions of opposite charges) will aggregate and form clusters that are different from a "random mixture". Imagine, if you choose an arbitrary ion with charge $z_c e$ at the center, you will see that many counterions will be attracted to the center and form a "cloud of charges" to "neutralize" $z_c e$. This "cloud" will disappear when you remove the central charge $z_c e$. Furthermore, a second "cloud" will form at the next neighborhood to counteract this first layer of clouds. These "clouds" are called "ion cospheres". The center ion then has an entourage of ion cospheres surrounding it. If you put a test charge e at distance r from center, it will interact with the center charge according to

Coulomb's law, plus it will also interact with all other ions in the cospheres. The mean electrostatic potential is define as the work required to bring against the charge $z_c e$ a test charge from $r = \infty$ to distance r. The total electrostatic potential is then the sum of the Coulomb interaction with $z_c e$ and all the cosphere interactions that the test charge will experience coming from $r = \infty$ to r. We call this energy the mean or **average electrostatic potential (AEP)**, $\Psi(r)$, which, as described above, is different from the bare Coulomb energy between the pair of charges $z_c e$ and e. Namely, it is not equal to the Coulomb interaction of two charged bodies in isolation. It is the averaged potential that a unit charge will experience in a solution of ions upon taking into account of the distributions of other ions in the solution. Debye and Hückel described this interaction by Poisson's equation. (iii) Debye and Hückel further made three more approximations in the course of analyses: (a) the distribution of ions in the cospheres obeys the *Boltzmann distribution*; (b) the resulting Poisson-Boltzmann equation can be *linearized*; and (c) the ions (cations and anions) have no volume, i.e. they are *point charges* (like a geometric point). We shall use these simplifications when solving the Poisson equation.

4.1. Solution of the Poisson Equation & the Debye Screened Potential

For the charge $z_c e$ at center together with its cospheres of ions, a unit test charge at radius r from the center will experience interactions with the center charge as well as the ions in the cospheres. Poisson's equation from electrostatics says

$$\nabla^2 \Psi_c(r) \;=\; -\frac{4\pi}{\varepsilon_m} \rho_c(r) \qquad (4.1.1)$$

where $\rho_c(r)$ is the charge density (Coulomb/m^3) at r from the center of charge $z_c e$. It captures all the ions distributed in a differential volume dr at r. $\Psi_c(r)$ is the average electrostatic potentials (AEP) experienced by this test charge at r from the center. The charge density, in statistical mechanics, can be expressed in terms of the pair correlation functions $g_{jc}(r)$ (also called the *radial distribution function*, a probability distribution function) as

$$\rho_c(r) = \sum_j z_j e \rho_j g_{jc}(r) \qquad (4.1.2)$$

where $g_{jc}(r)$ is the pair correlation of the type j ion surrounding the center ion c. The summation j is over all ionic species (Na^+, Cl^-, SO_4^{-2}, etc) situated at r. ρ_j is the number density of j (number of ions of species j per volume). $g_{jc}(r)$ according to Boltzmann is characterized by the energy of interaction. The *potential of mean force* (PMF) $W(r)$ is defined in terms of $g_{jc}(r)$.

$$g_{jc}(r) \equiv \exp[-\beta W(r)] \cong \exp[-\beta z_j e \Psi_c(r)] \qquad (4.1.3)$$

At the lowest order of approximation in the cluster expansion,[59] the pair correlation is given by the Boltzmann factor of the interaction energy (which is here identified as the AEP $\Psi_c(r)$). Note that $\beta = 1/(kT)$ is the reciprocal temperature, and $k=$ the Boltzmann constant; $T =$ absolute temperature. The second equality is an approximation to the potential of mean force by substituting the AEP for the PMF. Substitution into (4.1.1) gives

$$\nabla^2 \Psi_c(r) = -\frac{4\pi}{\varepsilon_m} \sum_j z_j e \rho_j \exp[-\beta z_j e \Psi_c(r)] \qquad (4.1.4)$$

This equation is the combination of the Poisson equation and the Boltzmann distribution, and is thus called the **Poisson-Boltzmann equation**. It has been well-studied in the literature. Debye-Hückel made further simplification by expanding the exponential term in (4.1.4) and retaining only the first order term (the linear term)

$$\exp[-\beta z_j e \Psi_c(r)] = 1 - \beta z_j e \Psi_c(r) + \dots \qquad (4.1.5)$$

Then

$$
\begin{aligned}
\nabla^2 \Psi_c(r) &= -\frac{4\pi}{\varepsilon_m} \sum_j z_j e \rho_j [1 - \beta z_j e \Psi_c(r)] = \\
&= -\frac{4\pi}{\varepsilon_m} \sum_j z_j e \rho_j + \frac{4\pi}{\varepsilon_m} \sum_j z_j e \rho_j \beta z_j e \Psi_c(r)]
\end{aligned} \qquad (4.1.6)
$$

By electroneutrality, the first term on the right is zero, $\Sigma z_j \rho_j = 0$.

$$\nabla^2 \Psi_c(r) = \kappa^2 \Psi_c(r) \qquad (4.1.7)$$

where we have defined

$$\kappa^2 \equiv \frac{4\pi e^2}{\varepsilon_m kT} \sum_j z_j^{\,2} \rho_j \qquad (4.1.8)$$

This κ is called the Debye-Hückel *inverse shielding length*. It has units of (length^{-1}). It can be written in terms of Bjerrum lengths, B_z (or B), one accounts for the valences, the other does not:

$$B_z \equiv \frac{e^2}{\varepsilon_m kT} |\, z_+ z_-\,| \qquad (4.1.9)$$

$$B \equiv \frac{e^2}{\varepsilon_m kT} \qquad 4.1.10)$$

These Bjerrum lengths have the unit of length. Using Coulombs and Joules for the variables, B will have the unit of meters. It is a measure of the "*coupling strength*" of the ionic solution, namely the circumstances that encourage strong ion-ion interaction or association. Thus B is large at low temperatures and low permittivity, both factors promote strong interaction and thus ion association. B_z in addition measures the types of ion-ion interaction. For 2-2 electrolytes ($|z_+|= 2 = |z_-|$) the Bjerrum length B_z is $|z_+ z_-| B = |2 \times 2| B = 4B$. The coupling is increased fourfold! In term of B, the Debye inverse length becomes

$$\kappa^2 \equiv 4\pi B \sum_j z_j^{\,2} \rho_j \qquad (4.1.11)$$

We define an "*ionic strength*" I (a common quantity used in electrochemistry):

$$I \equiv \frac{1}{2} \sum_j z_j^{\,2} \rho_j \qquad (4.1.12)$$

Then κ^2 can also be written as

$$\kappa^2 \equiv \frac{8\pi e^2 I}{\varepsilon_m kT} = 8\pi B I, \quad or \quad \kappa = \sqrt{8\pi B}\,\sqrt{I} \qquad (4.1.13)$$

To solve (4.1.7), we use the Laplacian in spherical coordinates (assuming symmetry in the θ and ϕ coordinates):

$$\frac{1}{r^2}\frac{d}{dr}\left[r^2\frac{d\Psi_c(r)}{dr}\right] = \kappa^2\Psi_c(r) \tag{4.1.14}$$

There are two linearly independent solutions:

$$\Psi_c(r) = \frac{C_1}{r}\exp(-\kappa r)+\frac{C_2}{r}\exp(+\kappa r) \tag{4.1.15}$$

The two constants C_1 and C_2 will be determined by two boundary conditions. First we say that the AEP is always finite, thus $C_2=0$. If not, the second term would grow without bound as $r \to \infty$. Second, we assume that the ions are all charged "points" without an excluded volume. As $r \to 0$, the interaction potential is dominated by the Coulomb potential (between the unit test charge at r and the center charge $z_c e$ at $r=0$) with negligible cosphere effects. Thus

$$C_1 \equiv \frac{z_c e}{\varepsilon_m} \tag{4.1.16}$$

$$\Psi_c(r) = \frac{z_c e}{\varepsilon_m r}\exp(-\kappa r) \tag{4.1.17}$$

Eq.(4.1.17) is called the *Debye screened potential*. It is the basis of the Debye-Hückel theory for all thermodynamic properties of electrolyte solutions. During the derivation, we have made several approximations: (i) the Boltzmann distribution of ions, (ii) the linearized exponential term, and (iii) the point charges for ions. These approximations enable us to obtain the screened potential (4.1.17). But they also diminish the accuracy of the Debye-Hückel theory. We shall describe further attempts to improve the Debye-Hückel theory in following chapters.

The form (4.1.17) is also called the *Yukawa potential* in statistical mechanics. We give below an example on how to calculate the Debye inverse length κ for a simple aqueous salt solution.

[Example 4.1] Sodium chloride is dissolved in water at 20°C with relative dielectric constant $D= 78.358$. For the molalities M given below,

find the molarities c with the given densities. Also calculate the Debye inverse length, κ.

Table 4.1.1. Summary of Calculations of the Inverse Debye Length, κ

M (molality)	d (g/cc)	c(molarity)	κ (1/A)
0.1	1.000115	0.09953	0.10384
0.2	1.00519	0.19871	0.14672
0.32	1.00998	0.31725	0.18538
0.50	1,01708	0.49409	0.23135
0.80	1.02865	0.78613	0.29182
0.12	1.04365	1.17023	0.35604

Answer: The conversion from molality M to molarity c can be achieved according to eq.(1.2.3). This is simply done and registered in the above table. (*Verify this!*) The evaluation of the Debye κ can be calculated from eq.(4.1.8) or (4.1.13).

$$\kappa^2 \equiv \frac{4\pi e^2}{\varepsilon_m kT}\sum_j z_j^{\,2}\rho_j = 8\pi BI$$

Let us list the relevant constants first:

$e = 1.60206E\text{-}19\ Coulomb$
$\varepsilon_m = D\varepsilon_o = (78.358)(111.2E\text{-}12)\ Coulomb^2/(N\ m^2)$
$k = 1.38054E\text{-}23\ J/K$
$T = 298.15K$
$z_+ = +1$
$z_- = -1$

For the number density ρ_j, we start from the molarity c. The conversion takes c moles per liter and convert to ρ molecules/Angstrom^3 via the Avogadro number (1 gmol = 6.022E+23 molecules). One liter = 1000 cc, and 1 cm = 10^8 Ångstroms. The dissociation of NaCl is

$$NaCl \underset{in\ water}{\rightarrow} Na^+ + Cl^-$$

For 1 gmol of *NaCl* dissolved, two gmols of ions form (one gmol of Na^+ and one gmol of Cl^-). The total number of moles of ions (cations plus anions) is 2 gmols. (We assume that the salt is completely ionized). For cation number density (number of cations per Angstrom^3)

$$\rho_+ = c_+ \frac{gmol}{liter} \frac{liter}{1000cm^3} \frac{cm^3}{(10^8 A)^3} \frac{6.022E23 \; ions}{gmol} \quad (4.1.18)$$

with a similar formula for ρ_-. For example, for $c_+ = 0.09953$ gmol Na$^+$/liter, we have

$$\rho_+ = 0.09953 \frac{gmol}{liter} \frac{liter}{1000cm^3} \frac{cm^3}{(10^8 A)^3} \frac{6.022E23 \; ions}{gmol} = 0.00005739 \frac{cations}{A^3}$$

ρ_- is the same $= 0.00005739$ anions/A^3. Now we have all the data we need. The Bjerrum length B is

$$B \equiv \frac{e^2}{\varepsilon_m kT} \quad (4.1.19)$$

If one substitutes the variables in m-k-s (meter-kilogram-second) unit system into eq.(4.1.19), B will have the unit of meters.

$$B \equiv \frac{e^2}{\varepsilon_m kT} = \frac{(1.60206)^2 E - 38}{(78.358)(111.2E - 12)(1.38054E - 23)(298.15)} = 7.15625E - 10 \, meter$$

We can now calculate κ^2.

$$\kappa^2 \equiv \frac{4\pi e^2}{\varepsilon_m kT} \sum_j z_j^2 \rho_j =$$

$$= \frac{4(3.14159)(1.60206)^2 E - 38 * (1.0E + 10 \; A/meter)}{(78.358)(111.2E - 12)(1.38054E - 23)(298.15)} \left[(+1)^2 0.00005739 + (-1)^2 0.00005739 \right] =$$

$$= 0.01078275 \; A^{-2}$$

Thus $\kappa = 0.10384$ (1/Å). Note that we have multiplied the ratio by a factor 10^{10} (1 meter $= 10^{10}$ Ångstroms), in order to convert the Bjerrum length B (meters) into Ångstroms. The resulting units of κ are reciprocal

Ångstroms (1/ Å). We can also calculate κ at the other molarities. The results are recorded in Table 4.1.1. (*Verify!*)

□

In the above table, we have calculated the Debye inverse length κ according to the correct definition (i.e. based on the molarity *c=mole/liter*). In practice, physical chemists also calculate κ_M directly from the molality *M= moles/kg water*, i.e. skipping molarity.

$$\kappa_M^2 \equiv \left[\frac{4\pi e^2}{\varepsilon_m kT}\sum_j z_j^2\left(M_j\frac{0.6022 d_0}{1000}\right)\right]\frac{1}{A^2}, \qquad \kappa_M \sim \overset{\circ}{A}^{-1} \qquad (4.1.20)$$

where d_0 (kg/liter) = pure solvent (water) density. Namely using M_+ (gmol cation/1 kg water) in place of c_+ (gmol cation/liter of solution) in eq.(4.1.18). For low salt concentrations, κ and κ_M differ little, because m_+ and c_+ are similar in magnitude. But the difference grows at higher concentrations. Thus one must define at the outset which κ is being used to avoid confusion.

4.2 The Debye-Hückel Thermodynamics

In this section we shall derive the thermodynamic properties of the electrolyte solutions based on the Debye-Hückel theory. We shall obtain the electrostatic (internal) energy, the Helmholtz free energy, the osmotic pressure, the Gibbs free energy, and the activity coefficients.

The internal energy we shall obtain is due to the Coulomb interaction only. In real solutions (the Lewis-Randall scale) there are contributions from other interactions (such as dipole-dipole (DD), dispersion (Disp), and hydrogen boning (HB) forces) to the internal energy. These are not included here. The reason, if we recall, is that we have used the McMillan-Mayer scale, where the water molecules have been "suppressed". This implies that we are in an *osmotic* system and the solvent effects are felt through the osmotic pressure. (There will be more detailed descriptions of the osmotic system and conversion to Lewis-Randall scale in Chapter 8.) In the *thermodynamic perturbation theory*,[59] we can write the Helmholtz free energies *A* as a sum of contributions from different molecular interactions; each contributed by one type of pair interaction:

$$A = A^{idg} + A^{\mathrm{Re}\,p} + A^{Disp} + A^{DD} + A^{HB} + A^{iD} + A^{ES}... \qquad (4.2.1a)$$

where the superscripts: *idg= ideal gas, Rep= repulsive, iD= ion-dipole;* and *ES = electrostatic.* From these individual Helmholtz free energies, we can write down the corresponding internal energies (via the Gibbs-Helmholtz relation).

$$U = U^{idg} + U^{Re\,p} + U^{Disp} + U^{DD} + U^{HB} + U^{iD} + U^{ES} \dots \qquad (4.2.1b)$$

The electrostatic internal energy U^{ES} of an ionic solution in the Debye-Hückel theory can be derived either (i) from a classical electrostatic consideration, or (ii) from the statistical mechanical formula. We shall use the electrostatics here. The statistical mechanical approach will be given in Chapter 6. In terms of dielectrics, we consider all the ions with their cospheres in solution as a collection of non-overlapping capacitors, each with a radius of $1/\kappa$ (κ being the Debye inverse length). Since we assume a very dilute solution, the ions are far apart and the cospheres are separated from one another and do not overlap. Let us review the dielectrics.

4.2.1. The capacitor

The capacitor in dielectrics is composed of a metal conductor or a pair of metal conductors separated by a thin dielectric material, so that they can be used to store charge (Coulomb). For example, a parallel-plate capacitor has two conducting plates, with a gap filled by an insulator (the dielectric). The plates may be few millimeters apart and do not touch each other. One plate has charge +Q, the other plate has charge –Q. The amount of charge is *proportional* to the voltage V_e across the two plates. The higher the voltage V_e, the more the charge Q will be, vice versa. Depending on the dielectric material (air, oil, or phenolics), the proportionality constant may vary. This coefficient is called the *capacitance C* of the capacitor.

$$C \equiv \frac{Q}{V_e} = \frac{Coulomb}{Volt} = Farad \qquad (4.2.2)$$

It has the unit of Farad (1 Coulomb per one volt = 1 Farad). To charge the parallel-plate capacitor from zero Coulomb to Q (+Q for one plate and –Q for the other plate), work W_e has to be done. The electric work

W_e is given by the voltage times Coulomb: $dW_e = V_e dQ$. Thus a capacitor with charge Q has work

$$W_e \equiv \int_0^Q V_e dQ = \int_0^Q \frac{Q}{C} dQ = \frac{Q^2}{2C} \qquad (4.2.3)$$

A capacitor does not have to be in the geometry of parallel plates. Other geometries are possible. The so-called self-capacitance for a spherical capacitor of radius R (for example the famous van de Graaff sphere generator) can be calculated from the formula

$$C \equiv \varepsilon_m R \qquad (4.2.4)$$

For instance, the earth has a capacitance of 0.71 millifarads.

4.2.2. Electrostatic energy

Now we return to ions with their cospheres (with radius $1/\kappa$) as spherical capacitors. Each of these capacitors has a capacitance $C = \varepsilon_m / \kappa$ (according to eq.(4.2.4)). The work W_i needed to charge one capacitor (of ion i) is (eq.(4.2.3))

$$W_i = \frac{Q^2}{2C} = \frac{(z_i e)^2}{2\varepsilon_m / \kappa} \qquad (4.2.5)$$

For N molecules of *NaCl* in water, we shall have N_+ cations and N_- anions in the solution. Note that $N_+ = N_- = N$, and $N_+ + N_- = 2N$. There are 2N cospheres for 2N ions. Thus the total work W_{tot} is the sum for all 2N cospheres (normalized by the volume V of the ionic solution)

$$\frac{W_{tot}}{V} = \sum_{i=1}^{2N} \frac{W_i}{V} = \sum_{i=1}^{2N} \frac{(z_i e)^2}{2V\varepsilon_m / \kappa} = \sum_{j=+,-} \frac{N(z_j e)^2}{2V\varepsilon_m / \kappa} = \sum_{j=+,-} \frac{\rho_j (z_j e)^2}{2\varepsilon_m / \kappa} = \frac{kT\kappa^3}{8\pi} \quad (4.2.6)$$

where we have changed from counting ions (index $i=1,2N$) to counting the ionic species (index $j = Na^+$, Cl^-, for instance). We have also applied the definition of the Debye κ. According to the first law of thermodynamics, the change in internal energy ΔU is the difference between heat Q' and work W'

$$\Delta U = Q'-W' \tag{4.2.7}$$

This energy change ΔU is the electrostatic energy, U^{ES}. As heat $Q'=0$,

$$\frac{U^{ES}}{V} = -\frac{W_{tot}}{V} = -\frac{kT\kappa^3}{8\pi} \tag{4.2.8}$$

This is the contribution to internal energy coming from electrostatic interactions in the ionic solution. With the internal energy, we can calculate other thermodynamic quantities according to well-known thermodynamic relations. We list theses exact relations below.

From internal energy to Helmholtz free energy: (at constant V):
(The Gibbs-Helmholtz equation)

$$d\frac{A}{T} = Ud\left(\frac{1}{T}\right) \tag{4.2.9}$$

From Helmholtz free energy to pressure (at constant T):

$$dA = -PdV \tag{4.2.10}$$

From Helmholtz free energy to chemical potential (at constant T and V):

$$\left(\frac{\partial A}{\partial n_i}\right)_{T,V,n_j} = \mu_i \tag{4.2.11}$$

Once we have the chemical potential, μ_i, we can obtain the activity coefficient γ_i from formulas in Chapter 2. We use eqs.(4.2.8) and (4.2.9) to derive the (electrostatic) Helmholtz free energy. The result is (*verify!*)

$$\frac{A^{ES}}{V} = -\frac{kT\kappa^3}{12\pi} \tag{4.2.12}$$

Then from (4.2.10), the (osmotic) pressure P^{osm} is obtained.

$$\frac{P^{osm}}{\rho_{tot}kT} = 1 - \frac{\kappa^3}{24\pi\rho_{tot}} \tag{4.2.13}$$

where $\rho_{tot} = \rho_+ + \rho_-$ is the sum of the number densities of the cations and anions. The ratio $P^{osm} /(\rho_{tot} kT)$ is also known as the osmotic coefficient, ϕ.

$$\phi \equiv \frac{P^{osm}}{\rho_{tot}kT} \qquad (4.2.14)$$

Note that this ϕ is defined in terms of the osmotic pressure, P^{osm}. In physical chemistry, a different osmotic coefficient "ϕ^a" is defined. This ϕ^a will be shown in fact to be the solvent activity instead! Next from its definition, the Gibbs free energy

$$G = A + PV \qquad (4.2.15)$$

We have (upon removing the ideal gas contribution, $\phi^{idg} = 1$)

$$\frac{G^{ES}}{V} = \frac{A^{ES}}{V} + \frac{P^{osm}V}{V} = -\frac{kT\kappa^3}{12\pi} - \frac{kT\kappa^3}{24\pi} = -\frac{kT\kappa^3}{8\pi} \qquad (4.2.16)$$

We note that the Gibbs free energy in the DH approach is the same as the electrostatic internal energy U^{ES}. (This is only valid in the Debye-Hückel theory.) Next we derive the chemical potential of ion i from the Helmholtz freee energy (note: not from the Gibbs free energy). We can do this by restricting the differentiation at constant T and V. Note also that the derivative of the Debye κ with n_j (the number of ions j) is obtained from the definition of κ,

$$\frac{\partial \kappa}{\partial n_j} = \frac{\partial}{\partial n_j} \sqrt{\frac{4\pi e^2}{\varepsilon_m kT} (\sum_{k=+,-} \rho_k z_k^{\,2})} =$$
$$= \frac{1}{2} \left[\frac{4\pi e^2}{\varepsilon_m kT} (\sum_{k=+,-} \rho_k z_k^{\,2}) \right]^{-1/2} \left(\frac{4\pi e^2}{\varepsilon_m kT} \right) \left(\frac{z_j^{\,2}}{V} \right) = \frac{1}{2\kappa} \left(\frac{4\pi e^2}{\varepsilon_m kT} \right) \left(\frac{z_j^{\,2}}{V} \right) \qquad (4.2.17)$$

Thus

$$\frac{\mu_j^{ES}}{V} = \frac{\partial(A^{ES}/V)}{\partial n_j} = \frac{\partial}{\partial n_j} \left(\frac{-kT\kappa^3}{12\pi} \right) = \left(\frac{-kT}{12\pi} \right) \frac{\partial(\kappa^3)}{\partial n_j} = \left(\frac{-kT}{12\pi} \right) 3\kappa^2 \frac{\partial \kappa}{\partial n_j}$$

$$\qquad (4.2.18)$$

Substituting (4.2.17) into (4.2.18)

$$\frac{\mu_j^{ES}}{kT} = -\frac{1}{2}\left(\frac{e^2 \kappa z_j^{\ 2}}{\varepsilon_m kT}\right) \tag{4.2.19}$$

Since the activity coefficient is related to the chemical potential via

$$\frac{\mu_j}{kT} = \ln\frac{q_{rv}\Lambda_j^{\ 3}}{kT} + \ln(f_j^{\ \infty} x_j \gamma_j^{\ d}) \tag{4.2.20}$$

The μ_j^{ES} is the excess part of (4.2.20). Thus γ_j is

$$\ln\gamma_j = -\frac{1}{2}\left(\frac{e^2 \kappa z_j^{\ 2}}{\varepsilon_m kT}\right) = -\frac{1}{2}B\kappa z_j^{\ 2} = -z_j^2 B^{3/2}\sqrt{2\pi I} \tag{4.2.21}$$

The mean activity coefficients $\gamma_\pm$ (2.15) is then

$$\ln\gamma_\pm = -\frac{1}{2}\left(\frac{e^2 \kappa}{\varepsilon_m kT}\right)\left|z_+ z_-\right| = -\frac{1}{2}\kappa B_z = -\left|z_+ z_-\right| B^{3/2}\sqrt{2\pi d} \tag{4.2.22}$$

where $|z_+ \ z_-|$ is the absolute value of the product $(z_+)(z_-)$. (See the example below for proof.) We notice the dependence on $(-\sqrt{I})$ shows up clearly in the mean activity coefficient!

[Example 4.2] Show that using the definition (2.15) of the mean activity coefficients yields the form eq.(4.2.22)!
Answer: The mean activity coefficient $\gamma_\pm$ was defined in (2.15)

$$\ln\gamma_\pm \equiv \frac{v_+ \ln\gamma_+ + v_- \ln\gamma_-}{v_+ + v_-}$$

Using (4.2.21) for the individual activity coefficients γ_j (j=+,−) we have the expression

$$z_{mean} \equiv \frac{v_+ z_+^{\ 2} + v_- z_-^{\ 2}}{v_+ + v_-} = \left|z_+ z_-\right| \tag{4.2.23}$$

We want to prove the equality above. We start with the electroneutrality condition (1.1.4)

$$v_+ z_+ + v_- z_- = 0 \qquad (1.1.4)$$

First, we square it, the result is still zero ($0^2 = 0$)

$$(v_+ z_+ + v_- z_-)^2 = 0 = v_+^2 z_+^2 + v_-^2 z_-^2 + 2v_+ z_+ v_- z_-, \quad \text{or}$$
$$v_+^2 z_+^2 + v_-^2 z_-^2 = -2v_+ v_- z_+ z_- \qquad (4.2.24)$$

Second, we multiply (1.1.4) by ($v_+ z_+ - v_- z_-$),

$$(v_+ z_+ + v_- z_-)(v_+ z_+ - v_- z_-) \quad = 0 = v_+^2 z_+^2 - v_-^2 z_-^2 \qquad (4.2.25)$$

Now we add and subtract (4.2.25) from (4.2.24) to get

$$2v_+^2 z_+^2 = -2v_+ v_- z_+ z_-, \qquad v_+ z_+^2 = -v_- z_+ z_- $$
$$2v_-^2 z_-^2 = -2v_+ v_- z_+ z_-, \qquad v_- z_-^2 = -v_+ z_+ z_- \qquad (4.2.26)$$

Adding the last two terms gives

$$v_+ z_+^2 + v_- z_-^2 = -(v_+ + v_-) z_+ z_-,$$
$$\frac{v_+ z_+^2 + v_- z_-^2}{v_+ + v_-} = -z_+ z_- = (z_+)(-z_-) = |z_+ z_-| \qquad [Q.E.D] \qquad (4.2.27)$$

□

Activity coefficients are an important part of the ionic solution properties. The Debye-Hückel theory gives the $\ln \gamma_\pm$ as a negative quantity. When we apply the definition of the *ionic strength I* from eq.(4.1.12)

$$\ln \gamma_\pm = -\left(\frac{e^3 \sqrt{2\pi}}{(\varepsilon_m kT)^{3/2}} \right) |z_+ z_-| \sqrt{I} = -A|z_+ z_-|\sqrt{I} \qquad (4.2.28)$$

where A is a constant as defined by (4.2.28).

The Debye-Hückel Limiting Law: The Debye-Hückel mean activity coefficient ln $\gamma_\pm$ is proportional to the negative square root of I as in eq.(4.2.28).

The next question is how well does the Debye-Hückel ln$\gamma_\pm$ perform for real solutions? We compare the DH ln $\gamma_\pm$ with experimental data on NaCl solutions at $25°C$. Figure 4.2.1 shows that the DH ln $\gamma_\pm$ is valid only at very dilute ionic strengths, $I < 0.001M$. This result is very disappointing.

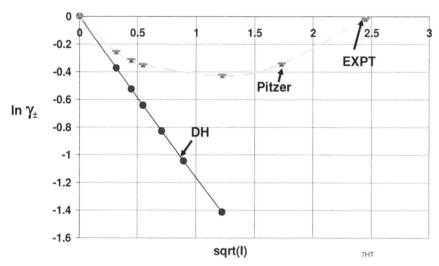

Figure 4.2.1. Performance of Debye- Hückel (DH) mean activity coefficient ln$\gamma_\pm$ (•) compared with experimental data (EXPT, ▬). Also shown are Pitzer's (next chapter) results (▲). (Sodium chloride at 20°C). DH is asymptotically valid at small ionic strength I.

However, it captures the infinite dilution behavior of ln $\gamma_\pm$, i.e. proportional to $(-\sqrt{I})$. Without the Debye-Hückel theory, this behavior is not possible to explain. The limiting law gives the infinite dilution behavior of electrolyte solutions, thanks to the Debye-Hückel theory.

In laboratory, one also uses a *practical ionic strength* I_m defined in terms of the molality, *M (gmol of ions/1 kg water)*, as

$$I_m \equiv \frac{1}{2}\sum_j z_j{}^2 M_j \qquad\qquad (4.2.29)$$

I_m also has the units of *(gmol/kg of water)*. This definition is very different from (4.1.12) for the rational I and will yield a different value. One should always verify the units.

Exercises:

4.1. Complete the Table 4.1.1 by calculating the Debye inverse length κ for the remainder molalities m > 0.1M.

4.2. For the entries in Table 4.1.1 find the electrostatic energy, the osmotic pressure, the Helmholtz free energy, and the Gibbs free energy using the Debye-Hückel theory.

4.3. For the entries in Table 4.1.1 find the individual and mean activity coefficients using the Debye-Hückel theory. Plot the ln γ± vs. the molality M.

4.4. Use eq.(4.2.9) to derive the Helmholtz free energy from the electrostatic energy.

4.5. In the McMillan-Mayer picture, the solvent is absent. However the solvent effect can be felt. It can be brought out by writing the Gibbs-Duhem relation (9.4.1) for a solvent a, and one cation species + and one anion species −:

$$x_a d\ln\gamma_a + x_+ d\ln\gamma_+ + x_- d\ln\gamma_- = \frac{V^E dP}{RT}$$

You can use the osmotic pressure dP^{osm} for dP. By inputting (4.2.21) for the activity coefficients, find the activity coefficient γ_a for the solvent a.

Chapter 5

Pitzer's Formulation for Electrolytes

Figure 4.2.1 shows that the Debye-Hückel $\ln \gamma_\pm$ is way off compared to the data (far too negative) and is inaccurate for practical calculations. Many attempts have been made to improve the accuracy. The methods of Güntelberg, Guggenheim, Davies, Bromley, and Meissner are such modifications. We cite only one of the more accurate ones that is much used in industry: the correlation due to Pitzer[83]. We remark that Pitzer's formulation is very accurate and correlated for many salts of interest (See Appendix I). However, it is semi-empirical with many fitted parameters. More rigorous approaches will be introduced in Chapter 7.

5.1. Pitzer's Correlation for Activity Coefficients

One of the modifications of widespread use is the one due to Pitzer.[83] His approach is based on the virial expansion of the activity coefficients. He retained the expansion up to the third virial coefficient.

$$\ln \gamma_\pm = -A_p|z_+z_-| + B_p m \frac{2\nu_+\nu_-}{\nu} + C_p m^2 \frac{2(\nu_+\nu_-)^{3/2}}{\nu} \qquad (5.1.1)$$

m is the molality. The virial coefficients A_p is based on the Debye theory; B_p, and C_p are fitted empirically to experimental data.

$$A_p = A\left[\frac{\sqrt{I_m}}{1+b\sqrt{I_m}} + \frac{2}{b}\ln(1+b\sqrt{I_m})\right], \quad where\ b = 1.2 \qquad (5.1.2)$$

where I_m is the practical ionic strength defined in (4.2.29) in terms of molality M (i.e. in units of *gmol/kg* of water). $b=1.2$ is a constant with units of $(I_m)^{-1/2}$. A is the coefficient similar to the Debye-Hückel (4.2.28), modified for use here:

$$A = \frac{1}{3}\left[\frac{e}{\sqrt{\varepsilon_m kT}}\right]^3 \sqrt{\frac{2\pi d_0 Avo}{1000}} \qquad (5.1.3)$$

where d_0 is the density (g/cc) for pure water at the system temperature (d_0 ~1.0 g/cc at the room temperature). *Avo* is the Avogadro number 6.022E+23. The division by 1000 is to change units from kg (in I_m) to grams. The Bjerrum length here should be expressed in *cm* (NOT meters!). The product $A(I_m)^{1/2}$ should be dimensionless; thus A has units of $(I_m)^{-1/2}$.

$$B_p = 2\beta_0 + \frac{2\beta_1}{\alpha^2 I_m}\left[1 - [1 + \alpha\sqrt{I_m} - 0.5\alpha^2 I_m]\exp(-\alpha\sqrt{I_m})\right], \quad where \; \alpha = 2.0 \qquad (5.1.4)$$

The constant α=2 has the units of $(I_m)^{-1/2}$.

$$C_p = \frac{3}{2}C^\phi \qquad (5.1.5)$$

β_0, β_1, and C^ϕ are parameters that are specific to the ionic species. (These are listed in Appendix I for many common salt species). To use the Pitzer's equations, the practical scale is employed. The units must be carefully identified. The molality is expressed as the number of gmols of ions in 1 kg of pure water.

For DH, the rational ionic strength I is expressed in number density (number of ions of type j per cc), ρ_j, see eq.(4.1.12). Let us use the cgs (centimeter-gram-second) unit system. The conversion between the number density and molarity c_j is

$$\rho_j = c_j\left[\frac{gmol}{liter}\right]\left[\frac{liter}{1000cc}\right]\left[\frac{6.022E + 23 \; ions}{gmol}\right] \qquad (5.1.6)$$

This ρ_j has units of (*number of ions j /cc*). We have used the Avogadro number (*Avo*) 6.022E+23 to convert the *gmol* to number of ions. The conversion between molarity c_j and molality m_j is eq.(1.2.3)

$$c_j = m_j\left[\frac{1000d_m}{1000 + \sum_{ions} n_{ion}W_{ion}}\right] \approx m_j d_0 \qquad (5.1.7)$$

We have made an approximation in the second equality that at low salt concentrations, $\Sigma_i n_i W_i$ (sum of moles of ion i multiplied by its molecular weight W_i) is much less than 1000, i.e. $\Sigma_i n_i W_i \ll 1000$. In this case the density of the mixture d_m is close to the density d_0 of pure water (at values close to 1 g/cc at room temperature). (5.1.6) can be written at low salt concentrations as

$$\rho_j = \left[\frac{m_j d_0 Avo}{1000}\right] \frac{number\ of\ ions}{cc} \tag{5.1.8}$$

ρ_j has units of *number of ions j /cc*. Thus the rational ionic strength I can be expressed at low concentrations as

$$I \approx \frac{1}{2}\left[\sum_{j=ions} z_j^2 \frac{m_j d_0 Avo}{1000}\right] \frac{number\ of\ ions}{cc} \tag{5.1.9}$$

On the other hand, the practical I_m (in molal units) is

$$I_m = \frac{1}{2}\left[\sum_{j=ions} m_j z_j^2\right] \frac{gmol}{kg\ of\ water} \tag{5.1.10}$$

Note that here I_m is expressed in units of *gmol/kg* of pure water. The best way to master the use of units in Pitzer's equations is to illustrate them by an example.

5.2. Example Calculation with Pitzer's Correlation

[Example 5.2.1] Calculate the mean activity coefficients $ln\gamma_\pm$ of the NaCl solution at the given concentrations (Table 4.1.1) using (1) Debye-Hückel formula, and (2) Pitzer's formula. Compare the calculated mean activity coefficients with the experimental data on $ln\ \gamma_\pm$.

Answer: We start with the first entry of Table 4.1.1 at *m= 0.1 gmol/kg of water*. The number densities ρ_j of ions are:

$$\rho_j = \left[\frac{m_j d_0 Avo}{1000}\right] \frac{number\ of\ ions}{cc} = \frac{(0.1)(1.0g/cc)(6.022E+23)}{1000} =$$

$$= 6.022E+19 \ \frac{number\ of\ j\ ions}{cc}$$

(Note: we do not use the previous units: *number of ions/Angstrom³*. We use ions/cc.) The rational ionic strength is

$$I \approx \frac{1}{2}\left[\sum_{j=ions} z_j^2 \frac{m_j d_0 Avo}{1000}\right]\frac{number\ of\ ions}{cc} = \frac{1}{2}\left[(1)^2 + (-1)^2]\frac{m_j d_0 Avo}{1000}\right] =$$

$$= \frac{1}{2}\left[(1)^2 + (-1)^2](6.022E + 19)\right] = 6.022E + 19 \quad \frac{number\ of\ ions}{cc}$$

The Debye formula (4.2.22):

$$\ln \gamma_\pm = -\left(\frac{e^3 \sqrt{2\pi}}{(\varepsilon_m kT)^{3/2}}\right)|z_+ z_-|\sqrt{I} = -|z_+ z_-|B^{3/2} \sqrt{2\pi I} \qquad (5.2.1)$$

Apply the definition of Bjerrum length

$$B \equiv \frac{e^2}{\varepsilon_m kT} = \frac{(1.60206)^2 E - 38}{(78.358)(111.2E - 12)(1.38054E - 23)(298.15)} = \quad 7.15625E - 10\ meter$$

Note that when expressed in *B*, the DH *lnγ*± is (4.2.22)

$$\ln \gamma_\pm = -B^{3/2}\sqrt{2\pi}|z_+ z_-|\sqrt{I} =$$

$$= -[(7.15625E - 10\ meter)(100cm/m)]^{3/2}\sqrt{2\pi}(1)\sqrt{6.022E + 19} = -0.37151$$

where we have converted the meters into cm and note that $|z_+ z_-| = |(1)(-1)| = 1$. <u>The DH *ln* $\gamma_\pm$ is −0.37151.</u>

The Pitzer formula (5.1.1):

The practical ionic strength I_m in Pitzer's equation is

$$I_m = \frac{1}{2}\left[\sum_{j=ions} m_j z_j^2\right] = 0.5*\left[0.1(1)^2 + 0.1(-1)^2\right] = 0.1 \quad \frac{gmol}{kg\ of\ water}$$

$$\sqrt{I_m} = \sqrt{0.1} = 0.3162 \quad \sqrt{\frac{gmol}{kg\ of\ water}}$$

The coefficient *A* is now

$$A = \frac{1}{3}\left[\frac{e}{\sqrt{\varepsilon_m kT}}\right]^3 \sqrt{\frac{2\pi d_0 Avo}{1000}} = \frac{1}{3}B^{3/2}\sqrt{\frac{2\pi d_0 Avo}{1000}} =$$

$$= \frac{1}{3}(7.15625E - 10\, meter * 100\, cm/m)^{1.5}\sqrt{2\pi}\sqrt{\frac{(1g/cc)(6.022E + 23)}{1000g/kg}} = 0.3925$$

The units of A is sqrt[*(kg of water)/gmol*], i.e. the reciprocal of sqrt(I_m). Then A_p is (Note that b= 1.2)

$$A_p = A\left[\frac{\sqrt{I_m}}{1+b\sqrt{I_m}} + \frac{2}{b}\ln(1+b\sqrt{I_m})\right] =$$

$$= (0.3925)*\left[\frac{0.3162}{1+1.2(0.3162)} + \frac{2}{1.2}\ln(1+2(0.3162))\right] = 0.30044$$

The Pitzer constants for *NaCl* are (from Pitzer's tables, Appendix I): β_0 =0.0765, β_1 =0.2664 and C^ϕ =0.00127. B_p is then (note that α =2)

$$B_p = 2\beta_0 + \frac{2\beta_1}{\alpha^2 I_m}\left[1 - (1 + \alpha\sqrt{I_m} - 0.5\alpha^2 I)\exp(-\alpha\sqrt{I_m})\right] =$$

$$= 2(0.0765) + \frac{2(0.2664)}{2^2(0.1)}\left[1 - \{1 + 2(0.3162) - 0.5[2^2(0.1)]\}\exp(-2(0.3162))\right] = 0.47129$$

C_p is simply: (1.5* C^ϕ = 0.001905). Thus the mean activity coefficient is

$$\ln\gamma_\pm = -A_p|z_+z_-| + B_p m\frac{2\nu_+\nu_-}{\nu} + C_p m^2\frac{2(\nu_+\nu_-)^{3/2}}{\nu} =$$

$$= -(0.30044)(1) + (0.47129)(0.1)\frac{2(1)(1)}{(1+1)} + (0.001905)(0.1)^2\frac{2((1)(1))^{3/2}}{(1+1)} = -0.25329$$

Pitzer's *$\ln\gamma_\pm$* = −0.25329. This is very different from the DH *$\ln\gamma_\pm$* = −0.37151 (47% in difference). The experimental data for *NaCl* at *M*= 0.1M is *$\ln\gamma_\pm$* = −0.2502. We can see that the Debye-Hückel *$\ln\gamma_\pm$* is not accurate, and Pitzer's correlation is accurate within 1.2%.   □

The accuracy of Pitzer's formula extends to the saturation point (~6 molal) for aqueous *NaCl* solution. Pitzer's equation is also accurate for a number of other salt solutions. Thus for engineering calculations, Pitzer's formulation can be used with a high degree of confidence For 2-2 type salts, the second virial coefficient B_p of Pitzer[84] will be modified.

5.3. Pitzer's Correlation for 2-2 Electrolytes

For 2-2 electrolyte solutions (such as $CuSO_4$), Pitzer[83] presented a modified formula for the mean activity coefficients. The modification is on the parameter B_p. The main formula is the same as eq.(5.1.1), i.e.

$$\ln \gamma_\pm = -A_p |z_+ z_-| + B_p^{2-2} m \frac{2\nu_+ \nu_-}{\nu} + C_p m^2 \frac{2(\nu_+ \nu_-)^{3/2}}{\nu} \tag{5.1.1}$$

The coefficients A_p and C_p are the same as before (eqs.(5.1.2 & 5.1.5)). The coefficient B_p for 2-2 electrolyte solutions is modified

$$B_p^{2-2} = 2\beta_0 + \frac{2\beta_1}{\alpha_1^2 I_m} \left[1 - [1 + \alpha_1 \sqrt{I_m} - 0.5\alpha_1^2 I_m] \exp(-\alpha_1 \sqrt{I_m}) \right] +$$
$$+ \frac{2\beta_2}{\alpha_2^2 I_m} \left[1 - [1 + \alpha_2 \sqrt{I_m} - 0.5\alpha_2^2 I_m] \exp(-\alpha_2 \sqrt{I_m}) \right] \quad where \; \alpha_1 = 1.4, \quad \alpha_2 = 12.0$$

$$\tag{5.3.1}$$

The parameters β_0, β_1, and β_2 are fitted to data and are listed in the Tables provided by Pitzer[84] (see Appendix I). The Tables in Appendix IV provide the experimental data for the mean activity coefficients $\gamma_\pm$ of the NaCl and KOH solutions at 25°C. The data are also plotted in Figures 5.3.1 and 5.3.2. We observe a distinct concentration dependence ($-\sqrt{I}$) at low molality, and $\gamma_\pm$ rises quickly after M> 2.0.

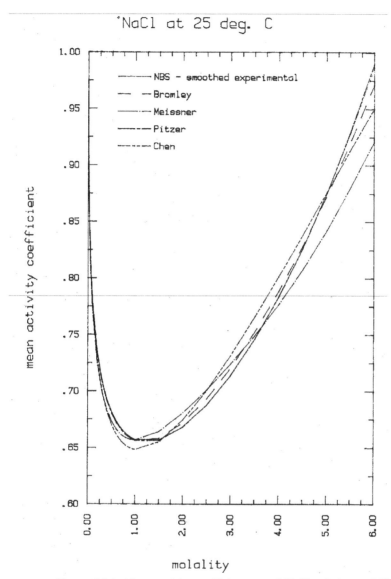

Figure 5.3.1. Mean activity coefficients $\gamma_\pm$ of NaCl solution at 25^oC. (NBS data). Other lines are from Bromley, Meisnner, and Chen.

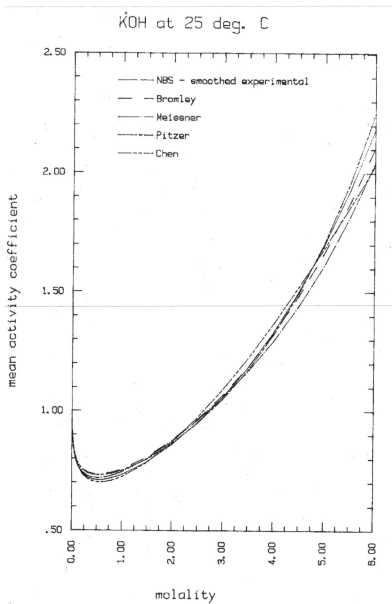

Figure 5.3.2. Mean activity coefficients $\gamma_\pm$ of KOH solution at 25°C. (NBS data). Other lines are from Bromley, Meisnner, and Chen.

Exercises:

5.1. Use Pitzer's equation to calculate the mean activity coefficients of NaCl at molalities listed in Table 5.1 for $0.001 < M < 6$. Pick ten points to make a graph.

5.2. Find the mean activity coefficients of salt $CuSO_4$ in water at $15°C$. (a)1.005M (density = 1.1573 g/cc). (b) 1.265M (density = 1.1965 g/cc). Use Pitzer's equation and Debye-Hückel theory. (Consult Appendix I for the parameters).

5.3. Given the density of aqueous solution of KCl = 1.1575 kg/liter at 3.9618 M ($10°C$), find the mean activity coefficients by Pitzer's method as well as Debye-Hückel theory.

5.4. From the thermodynamic relation

$$d\mu_i = \overline{V}_i dP \qquad \text{at constant T}$$

Find the osmotic pressure. Verify with the expressions from Pitzer in literature[80].

5.5. Find the mean activity coefficients of salt LiBr in water at $25°C$. (a) 1.2M (experimental value[65] = 0.834). (b) 15.0M (experimental value = 147.0). (c) 20.0M (experimental value = 486.0). Use Pitzer's equation and Debye-Hückel theory. (Consult Appendix I for the parameters).

Chapter 6

The Statistical Mechanics of Electrolytes

The molecular basis of thermodynamic properties is found in statistical mechanics. Any substance that is composed of "molecules" falls within the study of statistical mechanics. Recognizing that this book is not to teach statistical mechanics, we summarize the basics of this important scientific discipline below. Our purpose is to apply the principles of statistical mechanics to electrolyte solutions; to understand their behavior; to improve the results; and to advance, when possible, the state of art.

6.1. Basic Statistical Mechanics

In statistical mechanics, we consider all matter to be composed of molecules. A handful of matter will contain millions and billions of molecules. In fact, the number approaches Avogadro's number 6.022×10^{23}. We call a collection of material (e.g. gas in a cylinder, protein particles in a colloidal solution, or ions in a salt solution) an *N-body system*. N denotes the number of molecules or particles composing the system.

      The N-body system contains energy: the kinetic energy (KE) and the potential energy (PE). The kinetic energy is the energy due to the motion of the N bodies. It can be separated into three modes: translational kinetic energy (KE_t, due to translational motions), rotational kinetic energy (KE_r, due to rotations), and vibrational energy (KE_v, due to vibrations). On the other hand, the potential energy arises due to the "positions" or "relative positions" of the N molecules. On earth, we have the gravitational potential energy (thus *PE= mgh=(mass)(gravity g)(height)*). This is related to the position ("height" *h*) of the object above the earth that exerts a gravitation force "*–mg*". The relative positions of two molecules can also induce interaction forces between them, thus producing the interaction potential energy (such as the Coulomb energy between two charged particles that interact with the

Coulomb forces). The total energy (TE) of the N-body system is the sum of all the kinetic energies and the potential energies. The total energy is given the name the Hamiltonian, H_N. The Hamiltonian is just another name for the total energy (TE) of the N-body system:

$$H_N(r^N, p^N) = \sum_{i=1}^{N} \frac{p_i^2}{2m} + \sum_{1=i<}^{N} \sum_{j=2}^{N} u(r_i, r_j) \qquad (6.1.1)$$

where r^N is the N-vectors of the positions of the N molecules: (r_1, r_2, ..., r_N). p^N is the N-vectors of the translational momenta of the N molecules: (p_1, p_2, ..., p_N): $p_i = mv_i$, where m is the mass of one molecule, and v_i is the velocity (vector) of the *ith* molecule (i goes from 1 to N to count all N molecules). $u(i,j)$ is the pair interaction energy between a pair of molecules i and j. This interaction can be the Coulomb potential mentioned before, or the Lennard-Jones potential, or many other types of interaction potentials that describe the potential energy of the N-body system. Eq.(6.1.1) contains only the translational kinetic energies and the pair interaction energies. (For simplicity, we ignored the rotational and vibrational kinetic energies, and triplet, quadruplet, and higher n-body interactions). Due to the different types of interactions (potential energies), we have distinct properties for different chemical species. Water molecules interact with a water potential that is very different from, say, the potential of argon. Thus water has properties different from those of argon.

Once we have the Hamiltonian, the Boltzmann distribution law (a fundamental principle of statistical mechanics) says that in order for this N-body system to be at equilibrium, the molecules must distribute spatially and dynamically according to the *Boltzmann distribution*. The N-body probability density P_N is proportional to the *exponential* of the total energy (TE) or equivalently to the Hamiltonian

$$P_N(r^N, p^N) dr^N dp^N = \frac{1}{N! h^{3N}} \exp[-\beta H_N(r^N, p^N)] dr^N dp^N \qquad (6.1.2)$$

where $P_N \, dr^N \, dp^N$ is the (joint) N-body probability distribution that the N molecules have the configuration r^N and "commotion" p^N. β is the reciprocal temperature: $\beta = 1/(kT)$, k=Boltzmann constant, T= absolute temperature. h is Planck's constant. This Boltzmann distribution law (6.1.2) is the foundation of equilibrium statistical mechanics.

Statistical mechanics has developed over the last one and one half centuries and influenced many branches of physics. Its methods are multi-faceted and powerful. We classify into three or four categories. (i) The combinatorial analysis (as applied to traffic flow, the parking problem, polymer molecules on a lattice); (ii) partition function approach (e.g. ideal gas, Langmuir adsorption), (iii) integral equations of distribution functions (e.g., the Ornstein-Zernike equation, the BBGKY equation, etc.), and more recently (iv) molecular simulations: Monte Carlo (MC) simulation and molecular dynamics (MD) simulation. We assume some prior acquaintance with these methods. Interested reader ought to consult many of the specialized books in this field.[30,59,87] We discuss next only two approaches: (i) the *partition function*, and (ii) the *integral equations for molecular distributions*.

The Partition Function, Z_N

According to the Boltzmann distribution, the "normalization factor" Z_N of the probabilities is called the partition function

$$Z_N \equiv \frac{1}{N! h^{3N}} \int e^{-\beta H_N(r^N, p^N)} dr^N dp^N \qquad (6.1.3)$$

This is a multidimensional integral, over the 3N positions and 3N momenta. Only for extremely simple cases can it be evaluated analytically (such as for *ideal gas* molecules[59]). Eq. (6.1.3) is the partition function for a *canonical ensemble*[59] (a collection in phase space that represents constant N molecules, V volume, and T temperature).

The Ornstein-Zernike Equation for Distribution Functions

The Ornstein-Zernike equation is written out as a convolution integral of two types of correlation functions: $h(r)$ and $c(r)$

$$h_{jk}(r) - c_{jk}(r) = \sum_i \rho_i \int d\vec{s}\, h_{ji}(|\vec{r} - \vec{s}|) c_{ik}(s) \qquad (6.1.4)$$

where $h_{jk}(r)$ is the *total correlation* between a pair of molecules of species j and k, $c_{jk}(r)$ the *direct correlation*. ρ_i the number density of the molecules of species j (number of molecules per Å^3). ($i,j,k=$ labels of different species of molecules). These correlation functions $h_{jk}(r)$ and $c_{jk}(r)$ are different cross-sections of the probability distributions[59] for the molecules in the N-body system. They can be related to various thermodynamic properties[59] of the N-body system, such as the internal energy U, the pressure P, and the chemical potential μ.

6.2. Derivation of the Debye-Hückel Theory from Statistical Mechanics

It is possible to derive the Debye-Hückel (DH) theory from statistical mechanics. While the classical derivation above (Chapter 4) is sufficient, a derivation from statistical mechanics will show the simplifications (approximations) that the DH theory has made in arriving at the answers. To remove these approximations, we expect the statistical mechanics to do the job. Thus it will be instructive to examine how DH arises in the statistical mechanical formulation.

Let us consider, for simplicity, a single salt dissolved in water. In the *McMillan-Mayer picture*, the solvent is replaced by the dielectric continuum. The Ornstein-Zernike (OZ) equations for the ion species (cations and anions) are as shown in (6.1.4), where as before $h_{jk}(r)$ is the total correlation between ions j and k; $c_{jk}(r)$ their direct correlation; and ρ_i the number density of the ions (i,j,k = species of cation +, or anion −).

In order to reproduce the DH results, we assume low salt concentrations. As a consequence, the limiting behavior of the cluster expansions[30] of h_{jk} and c_{jk} at low densities will apply. In this approximation c_{jk} outside the hard core is approximated by the Coulomb potential, whereas h_{jk} is approximate by a linearization of the DH pair correlation function[8].

$$c_{jk}(r) \cong -\frac{u^{\pm}_{jk}(r)}{kT} \tag{6.2.1}$$

$$h_{jk}(r) \equiv g_{jk}(r) - 1 \cong \exp[-\beta z_j e \Psi_k(r)] - 1 \cong -\beta z_j e \Psi_k(r) \tag{6.2.2}$$

Note that the total correlation h_{jk} is defined as $h_{jk} = g_{jk} - 1$, where g_{jk} is the pair correlation (or radial distribution) given before. Here we have made two approximations. Eq.(6.2.2) proposes a linearization of an exponential (an approximation). It is valid only at high temperatures. The interaction potential $u^{\pm}_{jk}$ is the Coulomb electrostatic interaction for a pair of ions.

$$u^{\pm}{}_{jk}(r) \;=\; \frac{1}{\varepsilon_m}\frac{(z_j e)(z_k e)}{r} \tag{6.2.3}$$

Combining eqs.(6.2.1 & 6.2.2) and substituting into (6.1.4), the OZ equation becomes

$$h_{jk}(r) = -\beta\, z_j e \Psi_k(r) = -\frac{\beta z_j z_k e^2}{\varepsilon_m r} + \sum_i \beta^2 \rho_i \int d\vec{s}\; z_j e \Psi_i(|\vec{r}-\vec{s}|)\,\frac{z_i z_k e^2}{\varepsilon_m s} \tag{6.2.4}$$

We define a λ-function so that it is neutral and free of any electric charge

$$\Psi_i(r) \equiv \frac{z_i e}{\varepsilon_m}\lambda(r) \tag{6.2.5}$$

i.e. $\lambda(r)$ is a function to be found from the OZ eq.(6.2.4), which now reduces to a simpler form:

$$-\lambda(r) = -\frac{1}{r} + \sum_i \frac{\rho_i (z_i e)^2}{\varepsilon_m kT}\int d\vec{s}\;\frac{\lambda(|\vec{r}-\vec{s}|)}{s} = -\frac{1}{r} + \frac{\kappa^2}{4\pi}\int d\vec{s}\;\frac{\lambda(|\vec{r}-\vec{s}|)}{s} \tag{6.2.6}$$

We have used the definition of the Debye κ. The above integral is a convolution of two functions $\lambda(r)$ and $x(r) \equiv (1/r)$. Take the Fourier transform of (6.2.6) and apply the convolution theorem that changes a convolution in the r-space into a product in the q-space:

$$-\tilde{\lambda}(q) = -\tilde{x}(q) + \frac{\kappa^2}{4\pi}\tilde{\lambda}(q)\tilde{x}(q) \tag{6.2.7}$$

where tilde ($\sim$) denotes the three-dimensional Fourier transform. For any function $f(r)$ with polar and azimuthal symmetry, the three-dimensional Fourier transform is defined as

$$\tilde{f}(q) \equiv \frac{4\pi}{q}\int_0^\infty dr \sin(qr)\, r f(r) \tag{6.2.8}$$

and q is the reciprocal vector to r. Solving (6.2.7)

$$\tilde{\lambda}(q) = \frac{\tilde{x}(q)}{1 + \dfrac{\kappa^2}{4\pi}\tilde{x}(q)} \qquad (6.2.9)$$

The Fourier transform of $x(r) = (1/r)$ is known to be $(4\pi/q)$.

$$\tilde{x}(q) = \frac{4\pi}{q} \qquad (6.2.10)$$

Thus

$$\tilde{\lambda}(q) = \frac{4\pi}{\kappa^2 + q^2} \qquad (6.2.11)$$

The inverse transform of the above is (Note: the correspondence between the forward and backward Fourier transforms can be easily obtained from a Fourier transform table).

$$\lambda(r) = \frac{\exp(-\kappa r)}{r} \qquad (6.2.12)$$

Thus

$$\Psi_i(r) \equiv \frac{z_i e}{\varepsilon_m}\lambda(r) = \frac{z_i e}{\varepsilon_m r}\exp(-\kappa r) \qquad (6.2.13)$$

This is precisely the Debye screened potential obtained in eq.(4.1.17) by solving the Poisson equation. Wading through this statistical mechanical derivation of the Debye result serves at least two purposes: (i) determination of the conditions under which the DH theory is valid; and (ii) paving the way of future improvements of the DH theory if we can incorporate corrections that were missed in the original DH. It is clear that DH depends on simplifications on the correlations h_{jk} and c_{jk}, and these simplifications are more likely to be valid only at low concentrations and high temperatures. Another imprecision becomes clear: the OZ equation (6.2.4) was solved as if the correlation functions had the Coulombic form inside the hard core. The excluded volume effects (hard repulsion) were not properly accounted for.

6.3. Electrostatic Internal Energy from Statistical Mechanics

Statistical mechanics provides a means of obtaining the internal energy from (i) the interaction potentials $u_{ij}(r)$ between the molecules, and (ii) the pair correlation functions $g_{ij}(r)$. In general,[59] for a binary mixture of molecules of species i and j, the internal energy formula is

$$\frac{\beta U^{ES}}{V} = \frac{1}{2}\sum_{i,j}\rho_i\rho_j\int dr\, 4\pi r^2 \beta u_{ij}(r) g_{ij}(r) \qquad (6.3.1)$$

We have obtained the DH expression for $g_{ij}(r)$ as,

$$g_{ij}(r) \cong \exp[-\beta z_i e\Psi_j(r)] \cong 1 - \beta z_i e\Psi_j(r) = 1 - \frac{z_i z_j e^2}{\varepsilon_m kTr}e^{-\kappa r} \qquad (6.3.2)$$

Upon substitution, we obtain

$$\frac{U^{ES}}{VkT} = \frac{1}{2}\sum_{i,j}\rho_i\rho_j\int dr\, 4\pi r^2 \frac{z_i z_j e^2}{kTr}g_{ij}(r) =$$

$$= \frac{1}{2}\sum_{i,j}\rho_i\rho_j\int dr\, 4\pi r^2 \frac{z_i z_j e^2}{kTr}\left[1 - \frac{z_i z_j e^2}{\varepsilon_m kTr}e^{-\kappa r}\right] =$$

$$= \frac{1}{2}\sum_{i,j}\int dr\, 4\pi r\frac{e^2}{kT}[\rho_i\rho_j z_i z_j] - \frac{1}{2}\sum_{i,j}\rho_i\rho_j\int dr\, 4\pi \frac{z_i z_j e^2}{kT}\left[\frac{z_i z_j e^2}{\varepsilon_m kT}\right]e^{-\kappa r} =$$

$$= 0 - \frac{1}{2}\frac{\kappa^4}{4\pi}\int_{r=0}^{\infty}dr\, e^{-\kappa r}$$

$$(6.3.3)$$

Here we have used the electroneutrality condition and the definition of the Debye κ. The integration finally gives

$$\frac{U^{ES}}{VkT} = \frac{-\kappa^3}{8\pi} \qquad (6.3.4)$$

This proves the internal energy result for the Debye-Hückel theory. It is indeed the same as what we have obtained earlier using the classical electrostatic capacitors. Once the internal energy is at hand, all other thermodynamic quantities can be obtained by standard thermodynamic relations. In deriving (6.3.4), we have assumed that the permittivity is not a function of temperature. In actuality, it *is* a function of temperature. To correct for this dependence, the energy calculated from (6.3.4) should be corrected by the derivative

$$U^{ES,New} = U^{ES}\left[1+\frac{\partial \ln \varepsilon_m}{\partial \ln T}\right] \tag{6.3.5}$$

This is called the Born-Bjerrum correction[11]. In comparing the results to actual experimental data, one should include this correction.

We note that U^{ES} is the electrostatic part of the total internal energy. The total internal energy U^{TOT} in the Lewis-Randall scale for the ionic solutions should include not only the above U^{ES}, but as well the ideal gas part U^{Idg}, the dispersion interaction part U^{Disp}, the solvent internal energy U^{water}, and the energy due to solvent-ion interactions, U^{iw}. (See eq.(4.2.1b)).

6.4. The Dielectric Constants of Solvents

The dielectric constants for a number of industrial solvents are listed in Appendix III. It includes water and common organic solvents. We observe that the permittivity is a function of temperature. Normally, the higher the temperature, the lower is the permittivity. We discuss the evaluation of the permittivities of (i) pure solvents and (ii) mixtures of solvents.

6.4.1. The molecular-based formulas

One of the molecular formulas for permittivity is the Kirkwood equation. It depends on a Kirkwood factor g_K defined as

$$g_K \equiv 1 + \frac{\rho}{3}\int_0^\infty dr\, 4\pi r^2 h_\Delta(r) \tag{6.4.1}$$

where $h_\Delta(r)$ is the total correlation projected[59] onto the dipole-dipole inner product $\Delta=(\mathbf{L_1}\cdot\mathbf{L_2})$. $\mathbf{L_1}$ and $\mathbf{L_2}$ are the unit vectors of the dipoles 1 and 2. The coupling strength parameter γ_d in dipolar solvents is defined as

$$\gamma_d \equiv \frac{4\pi\mu^2\rho}{9kT} \tag{6.4.2}$$

where μ is the dipole moment, ρ the number density. γ_d is a measure of the coupling strength in dipolar liquids. Higher value means stronger binding interactions. The permittivity ε_m is given by the Kirkwood firmula[59]

$$\gamma_d g_K = \frac{(\varepsilon_m - 1)(2\varepsilon_m + 1)}{9\varepsilon_m} \qquad (Kirkwood\ formula) \tag{6.4.3}$$

Knowing the coupling strength γ_d and the Kirkwood factor g_K, the permittivity ε_m can be solved from (6.4.3). Two other popular formulas are the Clausius-Mossotti formula and the Onsager formula. The Onsager formula is obtained by setting $g_K = 1$

$$\gamma_d = \frac{(\varepsilon_m - 1)(2\varepsilon_m + 1)}{9\varepsilon_m} \qquad (Onsager\ formula) \tag{6.4.4}$$

The Clausius-Mossotti formula is

$$\gamma_d = \frac{(\varepsilon_m - 1)}{(\varepsilon_m + 2)} \qquad (Clausius\text{-}Mossotti) \tag{6.4.5}$$

Given the dipole moment μ, the density ρ, and the temperature, these equations are solved for the permittivity ε_m. To calculate with the Kirkwood formula, one needs also the total correlation function h_Δ. This quantity is obtained from a molecular theory.[59]

6.4.2. The permittivity of mixed solvents

For mixtures of several solvents, the dielectric constant is composition dependent. Let us define the polarization per volume, ω_i, for pure component i as

$$\omega_i \equiv \frac{(\varepsilon_i - 1)(2\varepsilon_i + 1)}{9\varepsilon_i} \qquad (6.4.6)$$

It is the RHS of eq.(6.4.3). Each pure solvent i will have a polarization ω_i. When n solvents are mixed, the *volume fraction averaged* ω_{mxt} is taken to be the mixture polarization.

$$\omega_{mxt} = \frac{\sum_i^n \omega_i x_i v_i}{\sum_i^n x_i v} = \frac{(\varepsilon_{mxt} - 1)(2\varepsilon_{mxt} + 1)}{9\varepsilon_{mxt}} \qquad (6.4.7)$$

where ε_{mxt} is the desired mixture permittivity, x_i is the mole fractions of i, and v_i is the molar volume of pure solvent i. Thus with known pure solvent permittivities and molar volumes, we can obtain the permittivity of the solvent mixture. Figure 6.4.1 shows the relative dielectric coefficient for the water-ethylene glycol mixture.

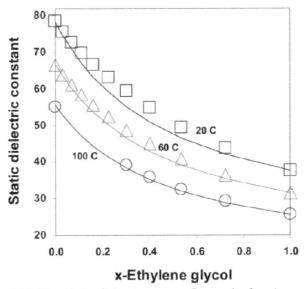

Figure 6.4.1. The relative dielectric constant, $D = \varepsilon_{mxt}/\varepsilon_0$, for mixtures of water and ethylene glycol (EG) as a function of mole fractions of EG. Symbols are experimental data. Linas are from the volume averaged formula (6.4.7). As temperature increases, D decreases.

We note that when salts are added, there is a dielectric decrement (decrease of dielectric constant) of the solution. For certain salt species, D can instead increase (such as for the onium salt: tetrabutylammonium perchlorate). Generally speaking, ion pairing enhances permittivity, while dissociated ions reduce ε_{mxt}. The behavior can also be more complicated than stated. We refer the reader to the references.[44,106]

Exercises:

6.1. What kind of electrostatic energy will one obtain if the pair correlation is not linearized?

$$g_{ij}(r) \cong \exp[-\beta z_i e \Psi_j(r)]$$

Substitute this in eq.(6.3.1). Evaluate it numerically for NaCl with $\kappa = 0.10384$ Å^{-1}. B= 7.16 Å. T= 25°C. ρ^+= 0.574 E-4 (cations/ Å^3). Compare it with eq.(6.3.4).

6.2. Find the Born-Bjerrum correction for water at 25°C. Correct the results from Problem 6.1 by this factor.

6.3. Compare the electrostatic energies of aqueous NaCl and $CuSO_4$ at T =25°C. Both are at 1.0M. (NaCl density = 1.036 g/cc. $CuSO_4$ density = 1.1573 g/cc). Use the Debye-Hückel theory.

6.4. Find the permittivity of a 50/50 mixture of water and methanol at 25°C.

Chapter 7

Ions as Charged Hard Spheres: The Mean Spherical Approach

We have seen that the Debye-Hückel results are inaccurate in describing concentrated salt solutions. Pitzer's equation with its virial coefficients does a remarkable job in giving accurate answers. However, Pitzer's formulation is largely empirical (fitting constants to salt data) and difficult to generalize to multi-solvent electrolytes. Since we have introduced the statistical mechanical approach, and we know what approximations were made in the Debye-Hückel theory, we can relax the conditions in DH to see if we can make improvements. One of the severest restrictions in DH is the point ion assumption: that ions, no matter how small (such as Li^+ ion with Pauling radius = 0.6 Å) or how big (as Br^- ion with Pauling radius = 1.95 Å), were assumed to be geometrical points, having no volume of exclusion (excluding other objects from invading the ion core). If we give the ions their proper volumes in the Ornstein-Zernike equations, what would happen to the physical properties of the ions? Would the result be better? This problem was solved[104] in 1970s in the *mean spherical approximation* (*MSA* or the mean spherical model). During the 1980s[37,38,56] and 1990s,[95,113] this approach has been shown to give more accurate results than the Debye-Hückel theory, and have a theoretical background that Pitzer's approach lacks. We shall adopt this approach as the basis for a number of industrial applications (i.e. acid gas treating and absorption refrigeration with electrolyte refrigerants).

7.1. The Mean Spherical Approach

The simplest way to impart ions with a volume is to endow them with a spherical hard core of diameter d (Figure 7.1.1). The shortcomings of the Debye-Hückel approach are the lack of exclusion at higher concentrations where ions can get in "contact" with each other and "feel" each other's corporeal presence, just like the case of real gas molecules versus the ideal gas molecules. Without exclusion, the pressure would be too low. Equipped with an excluded volume, the pressure would go up, and the mean activity coefficients will also increase. This trend correctly "cures" the excessively negative values of $ln\gamma_{\pm}$ in the DH theory at high ionic strengths.

In Figure 7.1.1 we depict the hard sphere ions Li^+ and Br^- coming from the lithium bromide in an aqueous solution. The lithium ion Li^+ is the smaller ion (the Pauling crystalline diameter is ~1.2Å). The bromide ion Br^- is the larger one (the Pauling crystalline diameter is ~3.9Å). These ions do not overlap due to their harsh repulsive cores. The presence of a hard core gives a more realistic representation of the physics at high concentrations.

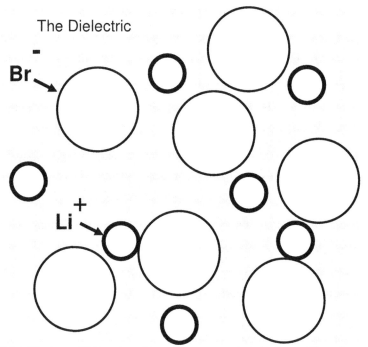

Figure 7.1.1. The ions of lithium bromide are given an excluded volume. The volumes are assumed to be spherical and the interaction between the volumes is hard core.

The interaction potential u_{jk} of the ions is the combination of: (i), a hard core for small distances that prevents overlap (with infinitely repulsive force), and (ii) the Coulomb interaction for distances larger than the excluded volume. We call this interaction the hard-sphere ions, or charged hard spheres.

$$u_{jk}(r) = \infty, \qquad if \quad r \le d_{jk}$$

$$u_{jk}(r) = \frac{z_j z_k e^2}{\varepsilon_m r}, \qquad if \quad r > d_{jk} \tag{7.1.1}$$

where d is the diameter of the spheres. For the j-j pair, the diameter will be d_{jj}; for the k-k pair, d_{kk}; and for the j-k pair, $d_{jk} = (d_{jj} + d_{kk})/2$. (We assume additive diameter here; i.e., the j-sphere and the k-sphere contact at a distance equal to the *arithmetic mean* of the two diameters d_{jj} and d_{kk}).

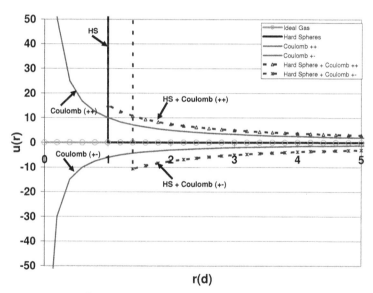

Figure 7.1.2. Depiction of the hard sphere and Coulomb interactions. Solid lines: Coulomb $++$ and $+ -$ interactions. Dotted lines: Coulomb interactions "decorated" with the hard sphere potentials (vertical lines).

In Figure 7.1.2 we contrast the pair interaction potentials for purely hard spheres with charged hard spheres. The x-y -coordinates are in arbitrary units. The baseline (with $u(r) = 0$) is the ideal gas interaction, which is identically zero! The hard sphere interaction is a vertical line, because it is infinitely repulsive. The Coulomb cation-cation interaction (point charges) is the positive ($1/r$) line. The Coulomb cation-anion interaction (point charges) is the negative ($-1/r$) line. They have no excluded volume, thus the curves enter the core region. If we decorate the Coulomb interactions by adding a hard sphere repulsion for $r < d_{jk}$, we would obtain the dotted lines, one positive (for hard sphere plus the Coulomb $++$ interaction), the other negative (for hard sphere plus the Coulomb $+ -$ interaction). The statistical mechanics is now assigned the task of solving for the probability correlations of the decorated potentials (eqs.(7.1.1)).

In the DH, the ion does not have volume. Thus the size of ion does not matter. The final equations contain only the quantities related to the charges, dielectric constant, density, and temperature. When the ions are endowed with a size, the effects of size are felt through the hard spheres that envelope these ions. Thus we choose the reference to be the neutral hard sphere mixture (i.e. mixture of neutral hard spheres with different sizes). Consider a sac of mixtures of golf balls and soccer balls. Although they all are spherical in shape, they are of different sizes. These are a neutral mixture of had spheres. The behavior of hard spheres is well-understood in literature. Now imagine these balls are charged with positive and negative charges. i.e., some are charged with q_1 Coulomb, and some are charged with q_2 Coulomb. Then you have a situation of the hard sphere ions. You would expect the statistical mechanics to give properties that depend not only on the valences, but also on the sizes: (1) sizes of ions d_1, d_2, etc.(2) charges of ions z_1, z_2, etc. and (3) the cross interaction between sizes and charges. This is precisely the case and all three factors are important.

The ***mean spherical approach*** is based on the recognition that (i) the pair correlation g_{jk} is known (exactly) ***inside the hard core***, i.e. equal to 0; and (ii) the direct correlation c_{jk} can be approximated ***outside the core*** by the long-range part of the pair potential.

$$g_{jk}(r) = 0, \qquad if \quad r \le d_{jk},$$

$$c_{jk}(r) = -\beta u_{jk}^{\pm}(r), \quad if \ r > d_{jk} \tag{7.1.2}$$

We shall not go into the details of the solution of the Ornstein-Zernike equations based on the above assumptions (7.1.2), knowing that to solve the equations, one would go into the Laplace space and make many algebraic manipulations.[104] We refer the reader to the references[10,104]. In literature, the *charged hard spheres* in the absence of the solvent molecules are called the **primitive model** (PM) of electrolytes. When, in addition, the ions have equal diameters ($d_{jj} = d_{kk}$), the electrolytes are called the **restricted primitive model** (RPM). When the solvent molecules are restored, we have the **non-primitive model**.

We summarize the results for the primitive model below without derivation. We first give the thermodynamic quantities. There are quite a number of factors that enter the expressions of the thermodynamic properties, such as P_n, Γ, η_n, Δ, Ω, etc. They will be defined following the description of the thermodynamic properties.

The Electrostatic Internal Energy, U^{ES}

$$\frac{U^{ES}}{VkT} = \frac{U^{Total}}{VkT} - \frac{U^{HS}}{VkT} = -B\Gamma\left[\sum_j \frac{\rho_j z_j^2}{1+\Gamma d_j} + \frac{\pi \Omega P_n^2}{2\Gamma\Delta}\right] \qquad (7.1.3)$$

when $P_n = 0$

$$\frac{U^{ES}}{VkT} \cong -B\Gamma\left[\sum_j \frac{\rho_j z_j^2}{1+\Gamma d_j}\right] \qquad (7.1.4)$$

The Electrostatic Helmholtz Free Energy, A^{ES}

$$\frac{A^{ES}}{VkT} = \frac{U^{HS}}{VkT} + \frac{\Gamma^3}{3\pi} \qquad (7.1.5)$$

The Osmotic Coefficient, ϕ^{ES}:

$$\phi^{ES} \equiv \frac{P^{osm}}{\rho kT} = \phi^{Total} - \phi^{HS} = -\frac{\Gamma^3}{3\pi\rho} - \frac{\pi B}{2\rho}\left[\frac{P_n}{\Delta}\right]^2 \qquad (7.1.6)$$

As $P_n = 0$

$$\phi^{ES} \cong -\frac{\Gamma^3}{3\pi\rho} \qquad (7.1.7)$$

The Mean Activity Coefficient, $\ln \gamma_{\pm}^{ES}$

$$\ln \gamma_{\pm}^{ES} = \ln \gamma_{\pm}^{Total} - \ln \gamma_{\pm}^{HS} = \frac{U^{ES}}{NkT} - \frac{\pi B}{2\rho}\left[\frac{P_n}{\Delta}\right]^2 \tag{7.1.8}$$

if $P_n = 0$

$$\ln \gamma_{\pm}^{ES} \cong \frac{U^{ES}}{NkT} \tag{7.1.9}$$

The Single-Ion Activity Coefficient, $\ln \gamma_j$

$$\ln \gamma_j^{ES} = -\frac{B\Gamma z_j^2}{1+\Gamma d_j} - \frac{\frac{\pi B}{2\Delta}z_j d_j P_n}{1+\Gamma d_j} - \frac{d_j P_n}{4\Delta}\left[\Gamma a_j + \frac{\pi^2 B}{3\Delta}d_j^2 P\right] - \frac{\pi B z_j}{6}\left[\sum_k \rho_k d_k^2 (\frac{2\Gamma a_k}{4\pi B}+\frac{z_k}{2})\right] \tag{7.1.10}$$

when $P_n = 0$

$$\ln \gamma_j^{ES} \cong -\frac{B\Gamma z_j^2}{1+\Gamma d_j} \tag{7.1.11}$$

The electrostatic (ES) part is the difference between the total property and the hard-sphere (HS) contribution. The hard sphere contribution can be obtained separately from the Mansoori-Leland-Carnahan-Starling (MLCS) hard sphere equation.[87] We cite some useful information from the mixtures of hard spheres.

The Equation of State for Mixtures of Hard Spheres[87]

$$\frac{P^{HS}}{kT} = \frac{6}{\pi}\left[\frac{\varsigma_0}{\Delta}+\frac{3\varsigma_1\varsigma_2}{\Delta^2}+\frac{3\varsigma_2^3}{\Delta^3}-\frac{\varsigma_3\varsigma_2^3}{\Delta^3}\right], \quad based \ on \ the \ MLCS \ equation \tag{7.1.12}$$

The Activity Coefficient for Hard Sphere Mixtures

$$\ln \gamma_j^{HS} = \frac{\pi d_j^3 P^{HS}}{6kT} - \ln \Delta + \left[\frac{3\varsigma_2 d_j}{\Delta}+\frac{3\varsigma_1 d_j^2}{\Delta}+\frac{9\varsigma_2^2 d_j^2}{2\Delta^2}\right] +$$

$$+ 3\left(\frac{\varsigma_2 d_j}{1-\Delta}\right)^2\left[\ln \Delta + \frac{\varsigma_3}{\Delta}-\frac{\varsigma_3^2}{2\Delta^2}\right] - \left(\frac{\varsigma_2 d_j}{1-\Delta}\right)^3\left[2\ln \Delta + \frac{\varsigma_3(2-\varsigma_3)}{\Delta}\right] \tag{7.1.13}$$

We have employed a number of factors P_n, Γ, η_n, Δ, Ω. Now we can define them below

The Size Factors, ζ_n and η_n

$$\eta_n \equiv \sum_j \rho(d_j)^n, \quad e.g., \quad \eta_2 \equiv \sum_j \rho(d_j)^2, \quad \eta_3 \equiv \sum_j \rho(d_j)^3 \qquad (7.1.14)$$

$$\zeta_n \equiv \frac{\pi}{6}\sum_j \rho(d_j)^n, \quad e.g., \quad \zeta_2 \equiv \frac{\pi}{6}\sum_j \rho(d_j)^2, \quad \zeta_3 \equiv \frac{\pi}{6}\sum_j \rho(d_j)^3 \qquad (7.1.15)$$

The Size Factor, Δ

$$\Delta \equiv 1 - \frac{\pi}{6}\sum_j \rho(d_j)^3 = 1 - \zeta_3 \qquad (7.1.16)$$

The Coupling of Size and Charge Factor, Ω

$$\Omega \equiv 1 + \frac{\pi}{2\Delta}\sum_j \frac{\rho_j(d_j)^3}{1+\Gamma d_j} \qquad (7.1.17)$$

The Coupling of Size and Charge Factor, P_n

$$P_n \equiv \frac{1}{\Omega}\sum_j \frac{\rho d_j z_j}{1+\Gamma d_j} \qquad (7.1.18)$$

Note that P_n will be zero when the hard core sizes are equal (by electroneutrality).

The Coupling of Size and Charge Factor, a_j

$$a_j \equiv \left(\frac{2\pi B}{\Gamma}\right)\left[\frac{z_j - \frac{\pi}{2\Delta}d_j^2 P_n}{1+\Gamma d_j}\right] \qquad (7.1.19)$$

The summation over j runs over all the ionic species: $j = Na^+$, Li^+, Cl^-, Br-, SO_4^{-2}, etc.

We see that there are geometric factors: η_n, Δ, and factors that couple the geometry with the electric charge: P_n, Ω. Note that ζ_3 is called the *packing fraction* (the volume actually occupied by the spheres). Δ is then the void fraction, or the fraction of free (empty) space in the entire container of volume V. The most important coupling parameter in MSA is Γ. Γ is given implicitly by

$$\Gamma^2 \equiv \pi B \left[\sum_j \rho_j \left(\frac{z_j - (\pi/2\Delta)d_j^2 P_n}{1 + \Gamma d_j} \right)^2 \right] \qquad (7.1.20)$$

Γ appears on both the left hand side and the right hand side of (7.1.20). Thus the equation should be solved by numerical method (or by iterations). This Γ has units of inverse length (1/Å). In fact, it is closely related to the Debye inverse shielding length κ. When the diameters, d_j → 0, the two are connected by (*Verify!*)

$$\kappa \cong 2\Gamma \qquad (7.1.21)$$

Namely, Γ is one half of the Debye κ. It also bears the same name *inverse shielding length*. In most practical calculations, P_n is a small number. When the ions are of same diameter, electroneutrality says $\Sigma \rho_j z_j$ =0, thus P_n =0. On first approximation, P_n can be set to zero. We have given the equations of $P_n = 0$ for the thermodynamic properties above. This simplifies the calculations considerably. The evaluation of Γ can be carried out by first calculating κ according to the Debye-Hückel formula. Next, we use ($\kappa/2$) as the initial guess for Γ on the right-hand side of (7.1.20) to calculate an approximate Γ on the LHS. Then we repeat the substitutions, putting the new Γ on the RHS to get a newer Γ. This shall accelerate the convergence of the numerical iterations. Normally, two to three iterations already give an answer Γ of enough accuracy.

[Example 7.1]: Find the mean activity coefficient of the NaCl solution at molality M= 1.5 molal and temperature T= 298.15K. The density is 1.01708 g/cc. Compare the MSA value with the experimental value of ln $\gamma_\pm$= −0.42068.
Answer: Due to the number of equations and variables to be calculated, it is better to use computer programming to solve this problem. A

Fortran program is appended to the end of the book for use. The output from this Fortran is given below.

OUTPUT FROM THE FORTRAN PROGRAM

```
==================================================
==================================================
    MEAN SPHERICAL APPROACH (MSA)
  Activity Coefficients of Charge Hard Spheres
    ----------------------------
    Based on Hoye-Blum 1978

    For Ionic Solutions of Aqueous NaCl

    By Lloyd LEE,        8/28/2007


==================================================
```

XXXXXXXXXXXXXXXXXXXXXXXXXXXXXXXXXXXXXXXXXXXXX

          For Debye Inverse Length, kappa (1/A)
            of NaCl Salt Solution at Temp=   298.15

This program uses the rational Scale:
Lengths in Angstrom, rho in no/A^3, Temp in K
The coupling parameter Pn is set to zero.

Relative Dielectric Const.=   78.358,permitt= 0.8713410E-08

Molal m=   1.5000, density g/cc= 1.01708, Molar c=   1.402547

Debye Inverse Shielding Length kappa= 0.3897550   1/A

XXXXXXXXXXXXXXXXXXXXXXXXXXXXXXXXXXXXXXXXXXXXXXX|
          For Hard Sphere Activity Coefficients InrHS

 Molarity:cmol= 1.402547
 rholon= 0.1689227E-02
 mole fractions ions x= 0.5000000   0.5000000
 diameters ions d= 2.800000   3.620000
 kappa 1/A 0.3897550
 zeta0= 0.8844772E-03
 zeta1= 0.2839172E-02
 zeta2= 0.9262421E-02
 zeta3= 0.3068690E-01
```
```

DELTA= 0.9693131
Term1= 0.3116762E-01
Term2= 0.8026750E-01
Term3= 0.6889138E-01
Term4= 0.3221436E-02
Term5= -0.2216679E-04
Term6= -0.6076702E-05
PHS= 0.2193084E-01

Activity Coefficient of HS, j= 1, lnrj= 0.2054505

Term1= 0.3116762E-01
Term2= 0.1037744
Term3= 0.1151505
Term4= 0.5384564E-02
Term5= -0.3705134E-04
Term6= -0.1313166E-04
PHS= 0.4739220E-01

Activity Coefficient of HS, j= 2, lnrj= 0.3028191

Mean Activity Coefficient from Hard Spheres, AvgHSlnr= 0.2541348

XXXXXXXXXXXXXXXXXXXXXXXXXXXXXXXXXXXXXXXXXXXXXX|
For MSA Activity Coefficients lnrMSA

Molarity:cmol= 1.402547
rholon= 0.1689227E-02
mole fractions ions x= 0.5000000   0.5000000
diameters ions d= 2.800000   3.620000   A
Bjerrum Length B= 7.156252   A
kappa 1/A= 0.3897550   A
eta0= 0.1689227E-02
eta1= 0.5422418E-02
eta2= 0.1768992E-01
eta3= 0.5860766E-01
DELTA= 0.9693131
GAMMA= 0.1203190
GAMMA= 0.1408489
GAMMA= 0.1345204
GAMMA= 0.1364090
OMEGA= 0.9347924

MSAlnr= -0.6799323

XXXXXXXXXXXXXXXXXXXXXXXXXXXXXXXXXXXXXXXXXXXXXX
Summary: Mean Activity Coefficients =MSAlnr+ HSlnr

for NaCl Solution at Temp=   298.15

Molality:m1= 1.500000
Molarity:cmol= 1.402547
mole fractions ions x= 0.5000000   0.5000000
diameters ions d= 2.800000   3.620000
Act.Coeff HSm lnrj= 0.2054505   0.3028191
Act.Coeff MSA lnr+- = -0.6799323

<<<<<<<<<<<<<<<<<<<<<<<<<<<<<<<<<<<<<<<<<
 Total Mean lnr =MSAlnr+MeanHSlnr= – 0.4257975
  >>>|

End Fortran Output

□

This $\ln\gamma_{\pm}$ can be compared with the experimental value of -0.4207 (an error of -1.2%). Note that we have used the ion sizes: for Na^+, $d_+ = 2.8\text{Å}$, and for Cl^-, $d_- = 3.62\text{Å}$ (the Pauling diameter). The cation diameter is not from Pauling's crystalline diameter (the latter should have been 1.90Å). Since in aqueous solutions, the cations are hydrated with 4 to 6 water molecules and have effectively a larger size than in the crystalline (solid) state. The effect of hydration will impact on the MSA results. In addition, correct physics says that salt addition affects the permittivity ε_m which depends on the amount of salt in the solution. The latter effect is neglected in the above calculation as a first approximation. The effect was actually absorbed into the determination of the hydration diameters.

Exercises:

7.1. For a hard sphere mixture (species 1+2) with diameters $d_{11} = 1$ (as unit length), $d_{22} = 1.5$. The packing fraction $\zeta_3 = 0.256$, find the pressure and activity coefficients of species 1 and 2.

7.2. For a primitive model of 2-2 electrolytes of charged hard spheres with the data in Problem 7.1, find the electrostatic energy, osmotic pressure, and mean activity coefficients. Let the relative dielectric constant $D = 78.358$, $T = 298.15K$.

7.3. Repeat the problem 7.2 by setting $P_n = 0$. Do you get different answers? How different?

7.4. Calculate the inverse shielding length Γ for $MgCl_2$ (a 2-1 electrolyte) at 298.15K and D=78.358 at 2.47M (density = 1.1720 g/cc). Compare it with the Debye κ. The Pauling crystalline radii for $Mg^{+2} = 0.65\text{Å}$, $Cl^{-1} = 1.81\text{Å}$.

7.5. Calculate the mean activity coefficient for $MgCl_2$ in the above problem. Compare with the experimental data $\gamma_\pm = 1.55$. Discuss the discrepancy, if any.

7.6. Find the osmotic pressure for sea water at 25°C using the MSA model. Sea water consists of about 0.5M of Na^+, and 0.55M of Cl^-. There are also 0.05M of Mg^{++}. The Pauling crystal radii for $Na^+ = 0.95\text{Å}$, for $Cl^- = 1.81\text{Å}$, and for $Mg^{++} = 0.65\text{Å}$. Water has relative dielectric constant of 78.358, and the Born-Bjerrum correction is -1.3679.

Chapter 8

The McMillan-Mayer and Lewis-Randall Scales

Starting with Debye and Hückel[23] in 1923, the electrolyte solutions have been studied with the solvent molecules rendered implicit and replaced by their permittivity (i.e. as a *dielectric continuum* with permittivity ε_m). This is known as the *McMillan-Mayer picture* or the **McMillan-Mayer scale**. This assumption greatly reduced the difficulty of treatment. However, when comparing with experimental data, the solvent has to be restored, namely account taken for the presence of the solvent molecules. All the formulas will have to be modified accordingly. To understand the difference in properties, we explain by a set of experiments from isobaric to isochoric to isoplethic. In fact, we shall describe four experiments that correspond to the various constrained thermodynamic conditions. For example, an experiment can be carried out at constant pressure (isobaric), or at constant volume (isochoric), or at constant solvent chemical potential, or at constant solvent/cosolvent ratio. Each route establishes a *scale*. Namely all thermodynamic properties in the particular scale are predetermined by the constraints (isobaric, isochoric, etc.) To change from one scale to another scale, the thermodynamic property (say osmotic pressure) will have to be converted according to known established thermodynamic relations.

The four scales we shall discuss are (1) The *Kirkwood-Buff scale* (KB); (2) The *McMillan-Mayer scale* (MM); (3) the *Lewis-Randall scale* (LR), and (4) the *Furter scale* (FS). All experiments will aim at preparing the same final salt solution–– but through different intermediate routes: starting from the pure solvent state (State *0*), continuing by adding small amounts of salt dn_s (States *k=1,2,3, ...*), eventually reaching the final salt concentration x_s^f (State *f*).

We begin with a *clean solution* (the pure solvent or a mixture of dielectric solvents, all free of salt). The solvents can be water, methanol, diethylene glycol, etc. The word *clean* means *no salt,* that salt concentration is zero. Solutions adulterated with salts are called *saline solutions.* The final solution shall have a salt mole fraction x_s^f. The solution may include many solvents as well as many salt species. We call these combinations multisolvent or multisalt. For simplicity, we consider at first a single salt s in a single solvent w.

Figure 8.1 depicts the Kirkwood-Buff experiment. A container of volume V^0 is first filled with the clean solvent. Constrained at *constant volume* V^0 and *constant temperature* T^0, small amounts of salt dn_s are added successively to the container fixed in volume until the final mole fraction x_s^f is reached. This procedure generates the pressures shown in Figure 8.5 along the curve "$0k$". We see that at the end point "k", the pressure of the system has reached a value P^k, which lies above the pressure P^0 of the initial clean solvent. It is understood that the pressure will rise rapidly as salt is being "squeezed" into a constant-volume container.

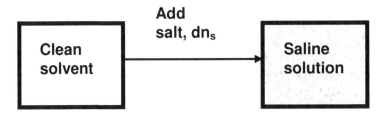

Figure 8.1. The constant volume cell in the Kirkwood-Buff charging process: Starting with a clean solvent, salt in small quantities is added to the container while maintaining a constant volume V^0 and constant temperature T^0.

Figure 8.2 shows the McMillan-Mayer experiment. It is an osmotic cell with two connected partitions, I and II. The two partitions are separated by only a *semi-permeable membrane*. The membrane is permeable to the solvent molecules, but not to the salt molecules. Initially, both Partition I and Partition II contain the clean solvent at identical conditions. Then salt is added to Partition II, but not to Partition I, while maintaining constant temperature T^0, constant volume V^0, and a constant chemical potentials μ_w^0 of the solvent (the chemical potential μ_w^{II} in Partition II is the same as μ_w^0 in Partition I at all times). The consequence is that the solvent chemical potentials are the same in both partitions despite the increase of salt in II. The solvent molecules can "swim" across the semi-permeable membrane and thus maintain the same chemical potentials. As salt is added to II, it at first tends to depress the chemical potential in Partition II, because μ_w is roughly proportional to $\ln(\,\rho x_w)$. As x_w drops, μ_w^{II} also decreases. In order to balance out μ_w^{0} on side I, the pressure in Partition II must increase. The pressure increase is maintained by a weight as shown, or by raising the liquid level in II (same phenomenon as the osmotic pull of the "*sap*" in trees). The driving force is thermodynamic, i.e. the Poynting effect ($d\mu_w = vdP$). The curve "**0m**" in Figure 8.5 shows the pressure trajectory of the MM experiment. The final pressure is P^1. The difference $P^1 - P^0$ is called the *osmotic pressure, P^{osm}*.

$$P^{osm} \equiv P^1 - P^0$$
(8.1)

P^1 rises less dramatically than P^k of the KB case because there is relief of solvent flow through the semi-permeable membrane. As a corollary to nature, this osmotic pressure is at the base of all plant life: it enables water to be "pumped" from the roots to the branches of the trees.

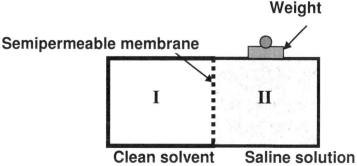

Figure 8.2. The osmotic cell in the McMillan-Mayer charging process. The left Partition I is filled with the clean solvent, w; the right Partition II is the saline solution,   Salts are added to Partition II while maintaining a constant solvent chemical potential μ_w^{0} in both Partitions. (Constant V^0 T^0 μ_w^{0}).

The third scale is the Lewis-Randall scale. It reflects the common laboratory practice. The process is carried out at constant pressure P^0 and constant temperature T^0. Line "**0L**" (the horizontal line) in Figure 8.5 indicates the course of this experiment. When data are taken for the thermodynamic functions along the three different trajectories described above, they refer to different state points and thus the data differ in values. The scale conversion is an attempt to convert one set of data on one trajectory to a set on the other trajectory.

Piston to keep at P°

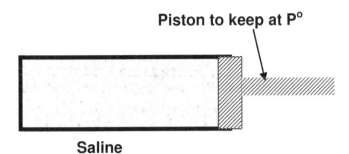

Saline

Figure 8.3. The piston-and-cylinder Cell: the Lewis-Randall charging process: Salt is added to the cell while keeping the pressure P^0 constant (with a movable piston to adjust volume) The temperature T^0 is also constant.

In the case of mixed solvents (i.e. there are more than one dielectric solvent, for example water the solvent + the amine monoethanolamine MEA as the cosolvent), there is also an experiment called the Furter chain–– chain of containers that contain increasing amounts of salt, while keeping the solvent ratio constant (e.g. weight of water/weight pf amine MEA = constant) in all containers. For more than two solvents, all solvent ratios are kept constant. The above are all experimental setups therein the solution thermodynamic properties can be measured. The data measured thereby are referred to as properties belonging to the particular scale.

Clean solvents   Saline solution   Final solution

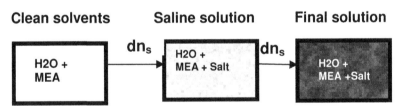

Figure 8.4. The constant solvent ratio cells: the Furter chain. The ratio of solvents water/MEA is kept constant form the first container to the last, while salt is added gradually until the final concentration is reached. (Constant T^0 and constant solvent ratio).

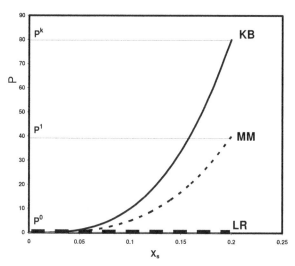

Figure 8.5. The pressure rise due to the addition of salts. Solid line = Kirkwood-Buff scale. P^k= final pressure of KB at final x_s. Dotted line = McMillan-Mayer scale. P^l= final pressure of MM. Flat dashed line = Lewis-Randall scale. P^0= the constant pressure in LR scale. P^{osm}= P^l-- P^0= osmotic pressure. (Pressure, P, is in arbitrary units. x_s = mole fraction of salt.)

In this chapter, we are concerned with the properties: the osmotic pressure and the activity coefficients. There are actually more than one way to convert from the MM scale to the LR scale, and from the MM scale to the KB scale: (i) by using thermodynamic arguments, and (ii) by using the Kirkwood-Buff[52] solution theory. While the thermodynamic conversion is useful in calculations, the Kirkwood-Buff fluctuation integrals[52] are theoretically more fundamental because they are related to the molecular correlations. Friedman[30] in 1960 and 1972 gave a benchmark discussion on the scale conversion between the McMillan-Mayer and Lewis-Randall frames. However, his formulations are (i) very complicated and thus difficult to implement, and (ii) not generalized to multisolvent electrolyte solutions. We present a simpler but equivalent procedure[65] below.

8.1. The Thermodynamic Route of Scale Conversion

One of the quantities that is much used in electrolyte chemistry is the osmotic coefficient. In practice there are at least two definitions. The first is the osmotic coefficient ϕ^{MM} that has been introduced in Chapter 4.

The other is the practical osmotic coefficient ϕ^{LR} that in actuality is the solvent activity (say, if water is the solvent, it is the water activity, a_w). The MM scale osmotic coefficient is

$$\phi^{osm} \equiv \frac{P^{osm}}{\rho RT} \equiv \phi^{MM} \qquad (8.1.1)$$

Recall that we have presented the osmotic pressure P^{osm} in the MM scale, eq.(7.1.6), therefore ϕ^{osm} refers to an osmotic pressure in a cell experiment in the MM scale. $\phi^{osm} = \phi^{MM}$ is the MM-scale osmotic coefficient. For a *single* solvent w plus a salt s, experimental chemists use a so-called *practical osmotic coefficient* ϕ^{LR} that is defined in terms of the **solvent activity** a_w, i.e.

$$\ln a_w \equiv -\phi^{LR} \frac{\sum\limits_{i=ions} m_i}{(1000/W_w)}, \qquad or \qquad \phi^{LR} \equiv -\ln a_w \frac{1000/W_w}{\sum\limits_{i=ions} m_i} \qquad (8.1.2)$$

where W_w is the molecular weight of the solvent (for water $W_w=18$), and summation on $i=ions$ runs over all ionic species. We make two remarks: (i) this practical osmotic coefficient ϕ^{LR} is "not" the osmotic pressure P^{osm} (at least not directly connected to it), but is defined in terms of the solvent activity a_w. (ii). ϕ^{LR} has the shortcoming that it is **not** defined in the environment of a multi-solvent solution because with many solvents (for example., water + methanol + ethanol), there is no singular solvent that can be the chosen to represent a_w. All three are eligible. *There is no unique choice!* Thus ϕ^{LR} is not defined for two or more solvents present simultaneously. It is only defined for single solvent electrolyte solutions. We shall call ϕ^{LR} the *solvent activity*. In contrast, the MM scale osmotic coefficient ϕ^{MM} has no ambiguities for the multi-solvent case and is well defined. In an osmotic cell, one can uniquely measure a physical pressure P^1 in the presence of all solvents, then P^{osm} is defined as $P^{osm} \equiv P^1 - P^0$. Next, we convert the solvent activity ϕ^{LR} to the mean activity coefficient $ln\ \gamma_{\pm}$ of the salt. This is done via the Bjerrum relation.

8.1.1. The Bjerrum relation

For a salt dissolved in water, (say, $NaCl$ (s) in water (w)), we can write down the Gibbs-Duhem relation for the activities a_w and a_s at constant temperature (T^0) and constant pressure (P^0).

$$n_w d \ln a_w + n_s d \ln a_s = 0 \qquad (8.1.3)$$

where n_w and n_s are number of moles of solvent and salt, respectively. We substitute a_w from eq.(8.1.2) to bring in the quantity ϕ^{LR}. When $NaCl$ dissolves in water, it dissociates into ions

$$
\begin{array}{ccccc}
NaCl & \rightarrow & Na^+ & + & Cl^- \\
m & & v_+ m & & v_- m
\end{array}
\qquad (8.1.4)
$$

On the basis of 1000 g of water and the molality unit, m moles of salt will have dissolved and produced $v_+ m$ moles of cations, and $v_- m$ moles of anions (in this case $v_+ = 1 = v_-$). Thus the summation Σm_i over all ions will give vm moles for all ions (where $v = v_+ + v_-$).

$$-\frac{1000g}{W_w} d \left(\phi^{LR} \frac{\sum\limits_{i=ions} m_i}{(1000/W_w)} \right) + vmd[\ln(\gamma_\pm) + \ln(m_\pm)] = 0 \qquad (8.1.5)$$

We note that the term $n_s dln(a_s)$ has been replaced with the help of the chemical equilibrium (8.1.4), i.e.

$\mu_s = v_+ \mu_+ + v_- \mu_-$,

$RT \ln a_s = v_+ RT \ln a_+ + v_- RT \ln a_-$,

$\ln a_s = v_+ \ln(\gamma_+ m_+) + v_- \ln(\gamma_- m_-) = v_+ \ln(\gamma_+) + v_- \ln(\gamma_-) + v_+ \ln(m_+) + v_- \ln(m_-) =$

$= v \ln(\gamma_\pm) + v \ln(m_\pm) =$

$= v \ln(\gamma_\pm) + v_+ \ln(v_+ m) + v_- \ln(v_- m) = v \ln(\gamma_\pm) + v \ln(m) + v_+ \ln(v_+) + v_- \ln(v_-) =$

$= v \ln(\gamma_\pm) + v \ln(v_\pm) + v \ln(m)$

$$(8.1.6)$$

where we have substituted the activities and activity coefficients for the chemical potentials. Next

$$-d(\phi^{LR}m) + md\ln(\gamma_{\pm}) + md\ln(m) = 0 \qquad (8.1.7)$$

Upon noting that

$$md[\ln(m)] = m\left(\frac{1}{m}\right)dm = dm \qquad (8.1.8)$$

we obtain

$$d[(1-\phi^{LR})m] + md\ln(\gamma_{\pm}) = 0$$

$$d\ln(\gamma_{\pm}) = -\frac{d[(1-\phi^{LR})m]}{m} = \frac{(\phi^{LR}-1)}{m}dm + d(\phi^{LR}-1) \qquad (8.1.9)$$

Integrating from zero salt concentration ($x_s = 0$, or $x_w = 1$) to final salt concentration x_s^f,

$$\ln(\gamma_{\pm}^{LR}) = (\phi^{LR}-1) + \int_{x_s=0}^{x_s^f}\frac{(\phi^{LR}-1)}{m}dm \qquad (8.1.10)$$

This is the *Bjerrum relation* that connects the practical osmotic coefficient ϕ^{LR} (namely the LR solvent activity) to the LR salt mean activity coefficient $\ln\gamma_{\pm}^{LR}$. The origin of this connection is the Gibbs-Duhem relation. Thus there should be no surprise that the *solvent activity* ϕ^{LR} is connected to the *salt* activity!

Next, we want to covert the MM osmotic coefficient to the LR solvent activity. Namely, we want to find a relation between ϕ^{MM} and ϕ^{LR}. We shall use the Poynting relation.

8.1.2. The Poynting relation

The Poynting correction is designed to determine the effects of pressure on the chemical potential. It is derived from a simple thermodynamic relation. Note that the Gibbs free energy G has the differential form for pure substances:

$$dG = VdP - SdT, \quad or$$
$$d\mu = vdP - sdT, \qquad (Constant\ T) \qquad (8.1.11)$$
$$d\mu = vdP,$$

The last equation is the Poynting correction. It says that under constant temperature T, the chemical potential differential is given by vdP (the specific volume $v=$ volume per mole times P). The increase in pressure will cause increase in the chemical potential (and vice versa). In mixtures, each species has its own correction. The molar volumes are replaced by the partial molar volumes, $\overline{V}_i$,

$$d\mu_i = \overline{V}_i dP \qquad (Constant\ T) \qquad (8.1.12)$$

Referring to Figure 8.2 the osmotic cell, we are dealing with two pressures: (i) P^0 at the initial *state 0*. The chemical potential of the solvent is μ_w^0. The molar volume is V^0. This initial state is represented by $(x_s^0 =0,\ x_w^0 =1;\ P^0,\ T^0)$. And (ii) P^f the final pressure. As salts are added, we eventually reach the final *state f*. The salt concentration is x_s^f, and the chemical potential of the solvent remains the same at μ_w^0. We represent this state as $(x_s^f,\ x_w^f;\ P^f,\ T^0)$. Since the initial chemical potential of the solvent μ_w^0 is the same as that of the final state

$$\mu_w(x_x^0, x_w^0, P^0, T^0)\ =\ \mu_w(x_x^f, x_w^f, P^f, T^0) \qquad (8.1.13)$$

Initial State 0 $\qquad\qquad$ *Final State f*

We interpose a third state: what does the final *state f* at $(x_s^f,\ x_w^f;\ P^f,\ T^0)$ have to do with a third state at a *low pressure state (LP)?* where the composition is the same as the final state but the pressure is at the initial value P^0 (i.e. going back to a low pressure state by decompressing the final pressure from P^f to P^0). This is where we apply the Poynting correction

$$\mu_w(x_x^f, x_w^f, P^f, T^0) - \mu_w(x_x^f, x_w^f, P^0, T^0) = \int_{P^0}^{P^f} \overline{V}_w(x_x^f, x_w^f, P, T^0)dP$$

$$\cong (P^f - P^0) < \overline{V}_w > = P^{osm} < \overline{V}_w >$$

(Final state f at P^f $\qquad$ *Low pressure state LP at P^0)* $\qquad (8.1.14)$

where we have applied the mean-value theorem on the partial molar volume $<Vw>$ (namely, $<Vw>$ is the partial molar volume of water at some intermediate pressure P' between P^0 and P^f). The pressure difference $P^f - P^0$ is precisely the osmotic pressure P^{osm} in the osmotic cell (Figure 8.2, the MM scale). When the clean solvent term $\mu_w(x_s^0,\ x_w^0;$ $P^0,\ T^0)$ is substituted into (8.1.14) for the chemical potential $\mu_w(x_s^f,\ x_w^f;$ $P^f,\ T^0)$, we obtain the activity of the solvent a_w

$$\mu_w(x_x^0,x_w^0,P^0,T^0) - \mu_w(x_x^f,x_w^f,P^0,T^0) = -RT \ln a_w(x_x^f,x_w^f,P^0,T^0) = P^{osm} <\overline{V_w}> \tag{8.1.16}$$

Upon applying the definition of the practical osmotic pressure (solvent activity) ϕ^{LR} of eq.(8.1.2),

$$-\ln a_w(x_x^f,x_w^f,P^0,T^0) = \phi^{LR}\frac{\nu m_s}{1000/W_w} = \frac{P^{osm} <\overline{V_w}>}{RT} = \phi^{MM}\rho<\overline{V_w}> \tag{8.1.17}$$

Rearrangements (see below) give

$$\phi^{LR} = \phi^{MM}(1 - c_s\overline{V_s}) \tag{8.1.18}$$

This is the *conversion formula between the LR and MM scales* for the osmotic coefficients ϕ^{LR} vs. ϕ^{MM}. Note that c_s is the molarity of the salt s, (moles of salt s per liter of the solution) and $\overline{V_s}$ is the partial molar volume of the salt s. For mixed salt solutions, we generalize to

$$\phi^{LR} = \phi^{MM}(1 - \sum_s c_s\overline{V_s}) \tag{8.1.19}$$

where s runs over the types of salts. Let us show the derivations from (8.1.17) to (8.1.18) below.

8.1.3. Proof of equation (8.1.18)

Let us take as basis 1000 g of water. Then from (8.1.17)

$$\phi^{LR} = \phi^{MM}\rho\overline{V_w}\frac{1000/W_w}{\nu m_s} = \phi^{MM}\frac{\nu m_s}{V}\overline{V_w}\frac{n_w}{\nu m_s} = \phi^{MM}\frac{n_w\overline{V_w}}{V} =$$

$$= \phi^{MM}\frac{n_w\overline{V_w}}{n_w\overline{V_w} + n_s\overline{V_s}} = \phi^{MM}\left(1 - \frac{n_s\overline{V_s}}{n_w\overline{V_w} + n_s\overline{V_s}}\right) = \phi^{MM}\left(1 - \frac{n_s\overline{V_s}}{V}\right) = \phi^{MM}\left(1 - c_s\overline{V_s}\right) \tag{8.1.20}$$

Note that we have used the exact condition that the total volume of the solution V is the sum of the moles of its components times their partial molar volumes $\overline{V}_j$

$$V = \sum_j n_j \overline{V}_j = n_w \overline{V}_w + n_s \overline{V}_s \qquad (8.1.21)$$

Also the molarity $c_s = n_s/V$. n_s is the number of moles of salt, and $n_w = 1000/W_w$ for water. We note that for small salt concentrations ($c_s \sim 0$), eq.(8.1.19) says that the two osmotic coefficients are almost the same. However, for high salt concentration, their difference becomes pronounced. Figure 8.1.1 shows the osmotic coefficients in the LR and MM scales for LiCl aqueous solutions up to 18 molal. ϕ^{LR} and ϕ^{MM} can differ up to 25% at M = 16 molal. The two osmotic coefficients are **not** the same (one is related to the osmotic pressure, the other to the solvent activity!).

Discussions-- Note that we have used the mean-value theorem to evaluate the integral in eq.(8.1.14). The mean partial molar volume $<V_w>$ absorbs the "compressibility effects". For incompressible liquids, this is not a bad approximation. (See also Simonin[94] 1996). The osmotic pressure can be very high (about 12 MPa at 5 molar and 25°C). To take compressibility into consideration, one can resort to the more rigorous formulation of Friedman.[30]

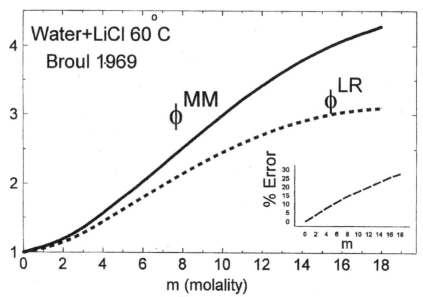

Figure 8.1.1. Comparison[65] of the osmotic coefficients from the Lewis-Randall scale ϕ^{LR} and that from the McMillan-Mayer scale ϕ^{MM}. Data are from Broul et al[14] (1969). The percent error between the two coefficients is plotted in the inset. The difference can reach 25% at 16 molal.

8.2. The Kirkwood-Buff Solution Theory in Scale Conversion

The Kirkwood-Buff solution theory[52] was proposed in 1951 as a bridge between the chemical potentials in classical (macroscopic) thermodynamics and the microscopic molecular distribution functions. It serves well here as a rigorous means of converting the Lewis-Randall quantities into the McMillan-Mayer quantities, and vice versa.

In this theory there are two matrices *A* and *B* of a Jacobian nature that have as elements chemical potential derivatives. *A* and *B* are inverse matrices of each other.

The A-matrix

The *A*-matrix consists of elements of the "compressibility" derivatives: partial derivatives of chemical potential μ_i of species i with respect to number of molecules of species j at constant volume V, constant temperature T, and *constant number $N_{!j}$* of species *not* of the j-type. (Note that $!j$ denotes *NOT j*).

$$A_{ij} \equiv \frac{V}{kT} \left(\frac{\partial \mu_i}{\partial N_j} \right)_{T,V,N!j} \tag{8.2.1}$$

The B-matrix

The **B**-matrix consists of elements that are inverses to the **A**-matrix. We have thus the "fluctuation" derivatives: partial derivatives of number N_j of molecules of species j with respect to chemical potential μ_k of species k at constant volume V, constant temperature T, and *constant chemical potential $\mu_{!k}$* of species *not* of the k-kind. (Note that $!k$ means *NOT k*).

$$B_{jk} \equiv \frac{kT}{V} \left(\frac{\partial N_j}{\partial \mu_k} \right)_{T,V,\mu!k} \tag{8.2.2}$$

It is an easy matter to show that **A** and **B** are inverse matrices of each other. These derivatives can be further shown to related to the molecular distribution functions: the total correlations $h_{jk}(r)$ and the direct correlations $c_{ij}(r)$ via

$$A_{ij} = \frac{V}{kT} \left(\frac{\partial \mu_i}{\partial N_j} \right)_{T,V,N!j} = \frac{\delta_{ij}}{\rho_i} - \int d\vec{r}\, c_{ij}(r) \equiv \frac{\delta_{ij}}{\rho_i} - C_{ij} \tag{8.2.3}$$

$$B_{jk} = \frac{kT}{V} \left(\frac{\partial N_j}{\partial \mu_k} \right)_{T,V,\mu!k} = \rho_j \delta_{jk} + \rho_j \rho_k \int d\vec{r}\, h_{jk}(r) \equiv \rho_j \delta_{jk} + \rho_j \rho_k G_{jk} \tag{8.2.4}$$

where we have defined the *fluctuation integrals* G_{jk} and *compressibility integrals* C_{ij} in terms of the total correlations $h_{jk}(r)$ and direct correlations $c_{ij}(r)$, respectively. Multiplication of the two matrices $\mathbf{A \cdot B} = \mathbf{I}$ (=unit matrix) gives

$$G_{ij} - C_{ij} = \sum_m G_{im} \rho_m C_{mj} \tag{8.2.5}$$

The Lewis-Randall scale

The primary variables in the LR scale are the pressure P and temperature T. Via mathematical manipulations (change of independent variables), we derive from the A-matrix

$$N_i \left(\frac{\partial \beta \mu_i}{\partial N_j} \right)^{LR}_{T,P,N\mathsf{!}j} = N_i \left(\frac{\partial \beta \mu_i}{\partial N_j} \right)^{KB}_{T,V,N\mathsf{!}j} - \frac{\rho_i \overline{V_i V_j}}{kTK_T} = \delta_{ij} - \rho_i C_{ij} - \frac{\rho_i \overline{V_i V_j}}{kTK_T} \qquad (8.2.6)$$

Since the activity coefficient can be obtained from the chemical potential

$$\mu_i = kT \ln(\gamma_i x_i f_i^\infty) + \mu_i^R \qquad (8.2.7)$$

$$N_i \left(\frac{\partial \ln \gamma_i}{\partial N_j} \right)^{LR}_{T,P,N\mathsf{!}j} = N_i \left(\frac{\partial \beta \mu_i}{\partial N_j} \right)^{KB}_{T,V,N\mathsf{!}j} - \delta_{ij} + x_i - \frac{\rho_i \overline{V_i V_j}}{kTK_T} \qquad (8.2.8)$$

where K_T is the isothermal compressibility

$$K_T \equiv -\frac{1}{V} \left(\frac{\partial V}{\partial P} \right)_T \qquad (8.2.9)$$

The McMillan-Mayer scale

The MM scale expressions can be derived from the B-matrix. Note that in the MM frame, the chemical potential of the solvent "α" is fixed during the salt-addition process. Thus μ_α = constant. In the following, if the index $i \neq \alpha$, then the derivatives are taken with fixed solvent chemical potential μ_α. This satisfies the MM frame. Now, from mathematics

$$\left(\frac{\partial \beta \mu_i}{\partial N_j} \right)^{MM}_{T,V,\mu\mathsf{!}i} = \frac{1}{\left(\dfrac{\partial N_j}{\partial \beta \mu_i} \right)^{KB}_{T,V,\mu\mathsf{!}i}} \qquad 8.2.10)$$

(Note that the index α is NOT to conflict with the index i. Or $i,j \neq \alpha$.). Applying the B-matrix

$$N_i \left(\frac{\partial \beta \mu_i}{\partial N_j} \right)^{MM}_{T,V,\mu'_i} = \frac{N_i}{\left(\dfrac{\partial N_j}{\partial \beta \mu_i} \right)^{KB}_{T,V,\mu'_i}} = \frac{1}{\delta_{ij} + \rho_j G_{ij}}, \qquad i \neq \alpha \qquad (8.2.11)$$

$$N_i \left(\frac{\partial \ln \gamma_i}{\partial N_j} \right)^{MM}_{T,V,\mu'_i} = \frac{N_i}{\left(\dfrac{\partial N_j}{\partial \beta \mu_i} \right)^{KB}_{T,V,\mu'_i}} - \delta_{ij} + x_i = \frac{1}{\delta_{ij} + \rho_j G_{ij}} - \delta_{ij} + x_i, \quad i \neq \alpha \qquad (8.2.12)$$

These formulas are "exact" formulas connecting the MM derivatives with the KB derivatives. We have assembled all the KB, LR, and MM formulas above in proximity, so they can be easily compared and converted. First of all, we see that the derivatives of the activities are different in all three frames. They do not share the same value! Quantities (partial molar volumes, isothermal compressibility, etc.) will have to be included to convert between them. In this approach, there is no direct connection between the MM and the LR quantities. The MM quantities must be converted to the KB quantities (8.2.12) first, then from KB to LR quantities (8.2.8). Offhand, if we have a molecular theory to generate the fluctuation integrals G_{ij} and C_{ij}, then it is straightforward to obtain quantities for all three scales. We leave the Furter scale to Chapter 9 on multi-solvent systems.

Exercises:

8.1. Use the Bjerrum relation (8.1.10) to find the solvent (water) activity ϕ^{LR} for NaCl solution at M = 0.01, 0.02, 0.03, 0.04, and 0.05. You can use Table 5.1 which gives the mean activity coefficients $\gamma_\pm$. (You need to invert the equation).

8.2. . Use the Bjerrum relation (8.1.10) to find the solvent (water) activity ϕ^{LR} for KOH solution at M = 2, 4, 6, 8, 10. You can use Table 5.1 which gives the mean activity coefficients $\gamma_\pm$. (You need to invert the equation).

8.3. Use eq.(8.1.18) to convert the ϕ^{LR} of NaCl obtained in 8.1 to the MM osmotic coefficients ϕ^{MM}. You may assume a constant $\overline{V}_s = 1$ cc/g.

8.4. Use eq.(8.1.18) to convert the ϕ^{LR} of KOH obtained in 8.2 to the MM osmotic coefficients ϕ^{MM}. You may assume a constant $\overline{V}_s = 0.89$ cc/g.

Chapter 9

Multi-Solvent Electrolyte Solutions: Setchenov's Salting-Out Principle

9.1. Introduction

Salt solutions containing two or more solvents (e.g., water, alcohols, amines, etc.) are important in the vapor liquid and solid-liquid equilibria that are found in many industrial processes: for example, bioseparation, absorption refrigeration, natural gas sweetening, hydrate inhibition, and dehydration. Thus much attention has been paid to the behavior of mixed-solvent systems.

Neutral liquid mixtures, such as water and ethanol exhibit an azeotrope behavior. Salt has large effects on this mixture. Adding enough salt species can "break" the azeotrope—displacing it or removing it entirely. Salt alters the vapor-liquid equilibrium of water and ethanol. The change depends on the salt used: some salts "like" water, or *hydrophilic*; other salts may dislike water, *hydrophobic*. Sometimes the affinity is a matter of degree: lithium bromide salt prefers water more than it prefers ethanol. Adding LiBr tends to drive ethanol out of the liquid phase and into the vapor phase. We call this behavior "*salting-out of ethanol*". On the hand, since LiBr has higher affinity for water, it will *salt-in* water (drawing water from the vapor phase into the liquid phase). This example illustrates the **salting-in** and **salting-out** behavior. These effects are of considerable industrial and theoretical importance. In separation processes such as extractive distillation, azeotrope distillation (Furter and Cook[32] 1967, Furter[33] 1972), extractive crystallization (Weingaertner et al.[109] 1991), biofluid processing (e.g. two-phase protein partitioning (Walters et al.[105] 1985)), or in geological formations (Harvey and Prausnitz[44] 1989) and in petroleum reservoirs, knowledge of salt effects is mandatory.

A number of studies have been proposed to quantify this behavior. The earliest is due to Setchenov[93] (1889). Later, the group of Furter[31-35] has carried out a considerable number of measurements on multi-solvent salt solutions, and put forth a Furter coefficient that links the relative volatility to the salt concentration. This was an advance over the Setchenov approach. To put the salting-out behavior on a more scientific basis, we adopt two approaches:[63-65] a thermodynamic approach and a molecular approach. The thermodynamic approach is based on (i) the Taylor expansion of the activity coefficients, and (ii) the Gibbs-Duhem relation. The molecular approach is based on the Kirkwood-Buff solution theory[52] as outlined in Chapter 8. These two approaches complement each other and form a sound theoretical basis for treating the salting-out behavior.

9.2. The Setchenov Principle

Setchenov[93] proposes a simple rule for the solubility of gas, g (for example methane), in a. saline solution (a mixture of a solvent and a salt, say, water with LiBr). At salt mole fraction x_s.

$$\ln\left(\frac{x_g^0}{x_g}\right) = k_s x_s \qquad (9.2.1)$$

where x_g is the mole fraction of the gas in the final saline solution (water + LiBr + methane) at salt concentration x_s. And x_g^0, is the solubility of gas in the clean solvent (water without salt). The coefficient k_s is called the *Setchenov constant*. Eq. (9.2.1) states that the logarithmic "decrease" of the gas solubility x_g depends "linearly" on the salt mole fraction in the solution. Depending on the sign of k_s, the solubility of the gas can either decrease (salting-out for positive k_s) or increase (salting-in for negative k_s). This rule seems to work well for a number of salt solutions and for gases such as He, H_2, N_2, O_2, argon, CH_4, and C_2H_6 (Pawlikowski and Prausnitz[81] 1983). Later, it was realized that this k_s is not a constant. It varies with x_s. Eq. (9.2.1) is nonlinear in salt concentrations beyond 5M, or mole fractions $x_s > 0.1$. A better correlation is needed. This is to be found in the Furter formulation.

9.3. The Furter Correlation

In a separate development, Johnson and Furter[49] employed the concept of relative volatility to determine the salt effects on mixed-solvent solutions. For example, for the mixture of two solvents water (a) +

methanol (b), we define the *relative volatility*, α_s between water and methanol in the presence of a salt LiBr (s) as

$$\alpha_s \equiv \left(\frac{y_a / x_a}{y_b / x_b} \right) \tag{9.3.1}$$

For a system of coexisting vapor and liquid phases, let x denote the mole fractions in the liquid phase, and y the mole fractions in the vapor phase. α_s is the ratio of the mole fractions y_a/x_a of the solvent a over the ratio y_b/x_b of the cosolvent b. When α_s is greater than unity, a is the more volatile solvent, and vice versa. In the absence of salt $x_s = 0$; we have only methanol + water. This salt-free solution is called a *clean solution*. The clean solution's relative volatility α_0 (at the same temperature) is similarly defined

$$\alpha_0 \equiv \left(\frac{y_a^0 / x_a^0}{y_b^0 / x_b^0} \right) \tag{9.3.2}$$

The superscript 0 denotes the clean solution properties. Furter says

$$\ln\left(\frac{\alpha_s}{\alpha_0} \right) \equiv k_F^0 x_s \tag{9.3.3}$$

The coefficient k_F^0 is called the Furter constant. This equation prescribes that: "the (logarithm of the) ratio of the saline α_s over the clean α_0 varies linearly with the salt content in the solution". This trend has again been shown to be successful for many mixed-solvent salt solutions (See Figure 9.3.1[32,33]).

A connection can be made between the Setchenov equation and the Furter equation. The Setchenov equation is a special case of the Furter equation when one of the cosolvent a is very dilute, say $x_a \to 0$, and solvent b is almost pure (both x_b and $x_b^0 \to 1.0$). Then eq.(9.3.3) becomes

$$\ln\left(\frac{\alpha_s}{\alpha_0} \right) = k_F^0 x_s = \ln\left(\frac{(y_a / x_a)(y_b^0 / x_b^0)}{(y_b / x_b)(y_a^0 / x_a^0)} \right) = \ln\left(\frac{x_a^0}{x_a} \right) = k_s x_s \tag{9.3.4}$$

Note that $y_b = y_b^0$, and $y_a = y_a^0$ (in a Furter chain experiment, both liquids—the saline solution and the clean solution share the same vapor phase). The Furter equation is of more general applicability. Figure 9.3.1 shows the linear variation of the relative volatility with addition of the salt potassium iodide *KI* in a water-ethanol binary solvent-cosolvent system. (Ethanol is at mole fraction $x_b^0 = 0.309$ in the clean solution). The linear relation continues to a salt mole fraction $x_s \sim 0.10$.

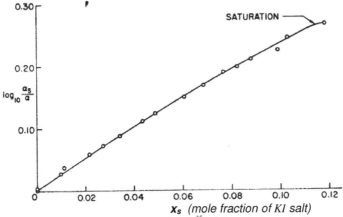

Figure 9.3.1. *Experimental data[35] of the relative volatility vs. the KI salt mole fraction x_s. The binary solvents are water (a) and ethanol (b). The ethanol mole fraction x_b in water is maintained at 0.309 for all data points.*

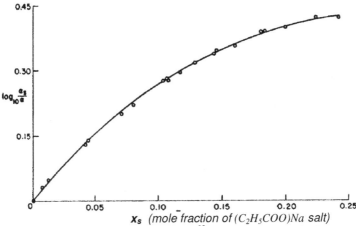

Figure 9.3.2. *Experimental data[35] of the relative volatility vs. the sodium acetate salt mole fraction x_s. The binary solvents are water (a) and ethanol (b). The ethanol mole fraction x_b in water is maintained at 0.309 for all data points.*

However, experimental data for other salts also show that Eq. (9.3.3) tends to fail for more concentrated salt solutions. Figure 9.3.2 shows the effect of the salt sodium acetate on the water-ethanol system. The ratio of the relative volatilities starts to deviate from linearity early at x_s about 0.05. Thus both Setchenov and Furter relations have their limitations. Since the above formulations are "empirically" obtained, we shall search for a more fundamental relation that is theoretically sound. We introduce the thermodynamic approach first. Afterwards, we examine the molecular theory.

9.4. The Taylor Expansion of the Activity Coefficients

In this section, we attempt to derive the Furter relation from the activity coefficients. Recall that at vapor-liquid phase equilibrium, the vapor phase fugacity f_i^V of component i is equal to the liquid phase fugacity f_i^L of the same component. Using the well-known γ-ϕ approach (γ for the liquid activity coefficient, and ϕ for the vapor fugacity coefficient):

$$f_i^L = f_i^V, \quad or \quad x_i \gamma_i f_i^{L0} = y_i P \phi_i^V \qquad (9.4.1)$$

where P is the system pressure, and f_i^{L0} is the pure liquid reference fugacity ($i=a,b$). Now as salt is added, the activity coefficients of solvents a and b will change. If we expand the activity coefficients in Taylor series in terms of the salt mole faction x_s (to first order in x_s) around the clean solution ($x_s=0$)

$$\ln \gamma_a = \ln \gamma_a^0 + \left[\frac{\partial \ln \gamma_a}{\partial x_s} \right]_0 x_s + ... \qquad (9.4.2)$$

Similarly for solvent b:

$$\ln \gamma_b = \ln \gamma_b^0 + \left[\frac{\partial \ln \gamma_b}{\partial x_s} \right]_0 x_s + ... \qquad (9.4.3)$$

Subtracting the two

$$\ln \gamma_a - \ln \gamma_b = \ln \gamma_a^0 - \ln \gamma_b^0 + \left[\frac{\partial \ln \gamma_a}{\partial x_s} - \frac{\partial \ln \gamma_b}{\partial x_s} \right]_0 x_s + ... \qquad (9.4.4)$$

Note that the activity derivatives are evaluated at zero salt concentration ($x_s = 0$, with subscript 0). The ratios of the activity coefficients by their definitions are

$$\left(\frac{\gamma_a}{\gamma_b}\right) = \left[\frac{(y_a P \phi_a^V)/(x_a f_a^{L0})}{(y_b P \phi_b^V)/(x_b f_b^{L0})}\right], \quad and \quad \left(\frac{\gamma_a^0}{\gamma_b^0}\right) = \left[\frac{(y_a^0 P \phi_a^V)/(x_a^{\ 0} f_a^{L0})}{(y_b^0 P \phi_b^V)/(x_b^{\ 0} f_b^{L0})}\right]$$

(9.4.5)

Substituting into (9.3.4)

$$\ln\left(\frac{(y_a/x_a)(y_b^0/x_b^0)}{(y_b/x_b)(y_a^0/x_a^0)}\right) = \left[\frac{\partial \ln \gamma_a}{\partial x_s} - \frac{\partial \ln \gamma_b}{\partial x_s}\right]_0 x_s = \ln\left(\frac{\alpha_s}{\alpha_0}\right) = k_F^0 x_s \qquad (9.4.6)$$

Comparing the second term with the last term we identify the Furter coefficient

$$k_F^0 = \left[\frac{\partial \ln \gamma_a}{\partial x_s} - \frac{\partial \ln \gamma_b}{\partial x_s}\right]_0, \quad (to\ first-order\ in\ x_s) \qquad (9.4.7)$$

We have obtained an interpretation of the Furter coefficient in terms of thermodynamically defined activity coefficients. For example, when the Wohl equation or the Wilson equation is used to represent the activity coefficients γ_i, we can carry out the differentiation to obtain the Furter k_F^0.

We have noted that at high x_s, the Furter equation becomes less accurate. This means that the first order terms in eqs.(9.4.2 & 9.4.3) are insufficient. If we include the second order terms, we shall have

$$k_F^1 = \left[\frac{\partial \ln \gamma_a}{\partial x_s}\right]_0 - \left[\frac{\partial \ln \gamma_b}{\partial x_s}\right]_0 + \frac{1}{2}\left[\frac{\partial^2 \ln \gamma_a}{\partial x_s^2} - \frac{\partial^2 \ln \gamma_b}{\partial x_s^2}\right]_0 x_s,$$

(to second-order in x_s) (9.4.8)

This gives an improved expression for the coefficient k_F.

9.5. The Gibbs-Duhem Relation for Multi-Solvent Systems

Since the electrolytes obey the same thermodynamic laws, the Gibbs-Duhem (GD) relation is obeyed by electrolyte solutions. We consider here a mixture of solvent a, cosolvent b, and salt s. The dissolved salt will dissociate into cations "+" and anions "−". We have four species in the mixture. The GD equation at constant temperature is

$$x_a d\ln\gamma_a + x_b d\ln\gamma_b + x_+ d\ln\gamma_+ + x_- d\ln\gamma_- = \frac{V^E dP}{RT} \qquad (9.5.1)$$

where V^E is the excess volume of mixing. And the mole fractions are defined as

$$x_a \equiv \frac{n_a}{n_a + n_b + n_+ + n_-}, \qquad x_+ \equiv \frac{n_+}{n_a + n_b + n_+ + n_-} \quad \text{(etc.)} \qquad (9.5.2)$$

(Note we are here solvent-explicit. The convention for explicit solvents was to have a primed quantity, x'. However, for simplicity, we drop the prime with the understanding that the solvent moles are counted in the total moles). If we differentiate with respect to the salt mole fraction, x_s, the GD becomes (upon ignoring the pressure term)

$$x_a \frac{\partial\ln\gamma_a}{\partial x_s} + x_b \frac{\partial\ln\gamma_b}{\partial x_s} + x_+ \frac{\partial\ln\gamma_+}{\partial x_s} + x_- \frac{\partial\ln\gamma_-}{\partial x_s} \approx 0 \qquad (9.5.3)$$

If we define an ion mole fraction, x_{ion}, (i.e. the total number of ions—sum of the numbers of cations and anions combined over the total number of moles) as

$$x_{ion} \equiv \frac{n_+ + n_-}{n_a + n_b + n_+ + n_-} = \frac{V_+ n_s + V_- n_s}{n_a + n_b + n_+ + n_-} = \frac{V n_s}{n_a + n_b + V n_s} \qquad (9.5.4)$$

(9.4.3) becomes

$$x_a \frac{\partial\ln\gamma_a}{\partial x_s} + x_b \frac{\partial\ln\gamma_b}{\partial x_s} + x_{ions} \frac{\partial\ln\gamma_\pm}{\partial x_s} \approx 0 \qquad (9.5.5)$$

Now if we take infinitely dilute salt concentration and multiply (9.4.7) by x_b^0 and add to (9.5.5) we can eliminate the cosolvent b-term:

$$(x_a^0 + x_b^0)\left[\frac{\partial\ln\gamma_a}{\partial x_s}\right]_0 = x_b^0 \frac{\partial(k_F^0 x_s)}{\partial x_s} - x_{ion} \frac{\partial\ln\gamma_\pm^\infty}{\partial x_s} \qquad (9.5.6)$$

Note that the superscript ∞ as in $\gamma_{\pm}^{\infty}$ indicates infinite dilution of salt for $\gamma_{\pm}$, while the superscript 0 on solvent γ_a^0 is pure solvent = infinitely dilute salt. This equation is a combination of the Gibbs-Duhem relation and the Furter relation. However, it is only applicable at infinitely dilute salt contents. Now if we generalize the Furter expansion (9.4.7) to any concentration x_s, then k_F becomes a function k_F^x of the salt mole fraction.

$$k_F^x(x_s) \equiv \left[\frac{\partial \ln \gamma_a}{\partial x_s} - \frac{\partial \ln \gamma_b}{\partial x_s} \right]_{x_s} \tag{9.5.7}$$

We can consider eq.(9.5.7) as a definition of a new function $k_F^x(x_s)$ valid at any x_s. It reduces to the Furter k_F^0 when $x_s=0$. Eq.(9.5.6) can then be generalized to a form valid for any concentration x_s

$$(x_a + x_b)\left[\frac{\partial \ln \gamma_a}{\partial x_s} \right]_{x_s} = x_b \frac{\partial (k_F^x x_s)}{\partial x_s} - x_{ion} \frac{\partial \ln \gamma_{\pm}}{\partial x_s} \tag{9.5.8}$$

A similar development can be made when we switch index from a to b,

$$(x_a + x_b)\left[\frac{\partial \ln \gamma_b}{\partial x_s} \right]_{x_s} = -x_a \frac{\partial (k_F^x x_s)}{\partial x_s} - x_{ion} \frac{\partial \ln \gamma_{\pm}}{\partial x_s} \tag{9.5.9}$$

Eqs.(9.5.8 & 9) are in a form that we can take advantage of in actual computations. If we know the mean activity coefficients $\gamma_{\pm}$ (from the MSA theory discussed above, say) and the Furter generalized $k_F(x_s)$ (from literature), then (9.5.8 & 9) allow us to integrate the right-hand side terms (all quantities are known) to obtain the activity coefficients γ_a. and γ_b on the left-hand side. This procedure was in fact carried out by Wu et al.[113,114] They applied (9.5.8 & 9) to the ternary mixture: water (*a*) + methanol (*b*) +LiCl salt (*s*). Experimental data on the vapor-liquid equilibria (VLE) were available from Broul et al.[14] The purpose was to use the combined Gibbs-Duhem-Furter theory to reproduce Broul's data (i.e. to check the utility of this combined approach). Wu fitted the Furter k_F^x to an equation:

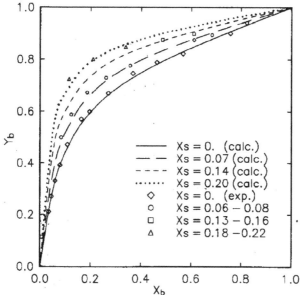

Figure 9.5.1. The x-y diagram for water(a)+methanol(b)+LiCl(s) system at 60°C (Data from Broul et al.[14] 1969). The vapor mole fraction y_b of methanol is plotted against liquid x_b as a function of the salt (LiCl) mole fraction x_s. The symbols are Broul's data. The lines are from the Gibbs-Duhem-Furter approach.

$$k_F^x(x_s) = 5 - (x_b^0 - 0.35)(1.588 - x_b^0)(1 + x_s)^{13.173} \qquad (9.5.10)$$

It is seen that k_F^x depends on the cosolvent (methanol) concentration x_b^0 (as it should), as well as on the salt concentration x_s. Once γ_a and γ_b are obtained, the VLE can be determined from (9.3.4) and (9.4.6). The results are shown in Figures 9.5.1 and 9.5.2. Figure 9.5.1 is an x-y diagram; Figure 9.5.2 is a P-x-y diagram. The lines are calculated from this combined GD-Furter theory. The symbols are from Broul's data. We can see a quantitative agreement between the theory and the data. The advantage of this combined approach is that the Gibbs-Duhem relation is automatically satisfied.

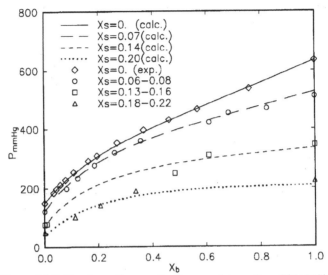

Figure 9.5.2. The P-x-y diagram for the water(a)+methanol(b)+LiCl(s) system at 60°C (Data from Broul et al.[14] 1969). Legend: same as in Fig. 9.1. Note that Broul's data are not isopleths (not at same salt compositions). Some variations in x_s exist in the data themselves.

9.6. The Kirkwood-Buff Solution Theory for Multi-Solvent Systems

We have alluded to the Kirkwood-Buff solution theory[52] in Chapter 8. Now we generalize it to the multi-solvent environment. We note that the expressions for the thermodynamic derivatives of the activity coefficients will change depending on which state variables are held fixed: $(T,V,n_i,...)$ or $(T,P, n_i, ...)$. For any quantity, Q, the mathematics of transformation of variables is (note $!j$ = not equal to j)

$$Q = \hat{Q}(T,V,\underline{N}) = \hat{Q}(T,\breve{V}(T,P),\underline{N}) = \breve{Q}(T,P,\underline{N}) \tag{9.6.1}$$

$$\left[\frac{\partial \breve{Q}}{\partial N_j}\right]_{T,P,!j} = \left[\frac{\partial \hat{Q}}{\partial N_j}\right]_{T,V,!j} + \left[\frac{\partial \hat{Q}}{\partial V}\right]_{T,N}\left[\frac{\partial \breve{V}}{\partial N_j}\right]_{T,P,!j} = \left[\frac{\partial \hat{Q}}{\partial N_j}\right]_{T,V,!j} + \left[\frac{\partial \hat{Q}}{\partial V}\right]_{T,N}\overline{V_j}$$
$$\tag{9.6.2}$$

Now let $Q = \beta\mu_i$,

$$\left[\frac{\partial \beta \bar{\mu}_i}{\partial N_j}\right]_{T,P,!j} = \left[\frac{\partial \beta \hat{\mu}_i}{\partial N_j}\right]_{T,V,!j} + \left[\frac{\partial \beta \hat{\mu}_i}{\partial V}\right]_{T,N} \overline{V}_j \qquad (9.6.3)$$

The Kirkwood-Buff solution theory says that

$$N\left[\frac{\partial \beta \hat{\mu}_i}{\partial N_j}\right]_{T,V,!j} = \frac{\delta_{ij}}{x_i} - \rho C_{ij}, \quad where \; C_{ij} \equiv \int c_{ij}(r)d\vec{r} \qquad (9.6.4)$$

From thermodynamics we know that $d\beta\mu_i = \beta\overline{V}_i \, dP$ at constant T ($\overline{V}_i$ is the partial molar volume of i):

$$N\left[\frac{\partial \beta \hat{\mu}_i}{\partial V}\right]_{T,N} = -\frac{\rho \overline{V}_i}{kTK_T}, \quad where \; K_T \equiv -\frac{1}{V}\left[\frac{\partial V}{\partial P}\right]_{T,N} \qquad (9.6.5)$$

K_T being the isothermal compressibility. Thus

$$N\left[\frac{\partial \beta \bar{\mu}_i}{\partial N_j}\right]_{T,P,!j} = N\left[\frac{\partial \beta \hat{\mu}_i}{\partial N_j}\right]_{T,V,!j} - \frac{\rho \overline{V}_i V_j}{kTK_T} = \frac{\delta_{ij}}{x_i} - \rho C_{ij} - \frac{\rho \overline{V}_i V_j}{kTK_T} \qquad (9.6.6)$$

Eq.(9.6.6) is the relation between the Lewis-Randall derivative (fixed T,P) and the Kirkwood-Buff derivative (fixed T,V) as discussed earlier. Furthermore, from the definition of the activity coefficient γ_i

$$\beta \mu_i = \beta \mu_i^* + \ln(x_i) + \ln(\gamma_i) + \ln(f_i^0) \qquad (9.6.7)$$

$$N\left[\frac{\partial \beta \mu_i}{\partial N_j}\right] = N\left[\frac{\partial \ln x_i}{\partial N_j}\right] + N\left[\frac{\partial \ln \gamma_i}{\partial N_j}\right] = \frac{\delta_{ij}}{x_i} - 1 + N\left[\frac{\partial \ln \gamma_i}{\partial N_j}\right] \qquad (9.6.8)$$

The above relation is valid at any fixed variables (at T, P or T, V). Substitution into (9.6.6) gives the derivatives of the activity coefficients

$$N\left[\frac{\partial \ln \gamma_i}{\partial N_j}\right]_{T,P,!j} = 1 - \rho C_{ij} - \frac{\rho \overline{V}_i V_j}{kTK_T} \qquad (9.6.9)$$

whereas from (9.6.4)

$$N\left[\frac{\partial \ln \gamma_i}{\partial N_j}\right]_{T,V,!j} = 1 - \rho C_{ij} \qquad (9.6.10)$$

Since in laboratory, we are under the Lewis-Randall scale: we keep T and P constant during the experiment. Thus eq.(9.5.9) is applied. The Furter relation is then

$$k_F^x(1-x_s) + x_s N\left[\frac{\partial k_F^x}{\partial N_s}\right] = (1 - \rho C_{as}) - (1 - \rho C_{bs}) - \frac{\rho \overline{V}_s(\overline{V}_a - \overline{V}_b)}{kTK_T} \qquad (9.6.11)$$

At low salt concentrations, $x_s \to 0$

$$k_F^0 \approx \rho(C_{bs} - C_{as}) - \frac{\rho \overline{V}_s(\overline{V}_a - \overline{V}_b)}{kTK_T} \qquad (9.6.12)$$

We have arrived at an expression of the Furter k_F^0 in terms of molecular quantities C_{ij} (plus partial molar volumes). These expressions are exact.

We have examined above a binary-solvent salt solution (two solvents and one salt). The formalism can be easily generalized to systems with n solvents and m salts ($n,m > 2$). We shall investigate these multi-solvent systems in the Chapter 13 on acid gas removal in natural processing.

Remarks:

We have presented two complete theories for mixed solvent electrolytes: the Gibbs-Duhem-Furter theory and the Kirkwood-Buff theory. Both are applicable to general multi-solvent systems. These approaches should and have supplanted the so-called *pseudo-solvent* methods in electrolyte solutions. The pseudo-solvent is a *hypothetical pure* solvent that has the same properties as the mixture (such as dielectric constant, density, etc.) It is used to avoid the complications of salting-out and salting-in behavior.

Exercises:

9.1. From the partial volume data of water and methanol below,[63] estimate the Kirkwood-Buff compressibility C-integrals C_{as}, C_{bs} and K_T from eq.(9.5.12). You may use eq.(9.4.10) at $x_s=0$.

Table 9E.1: The partial molar volumes* of methanol (a)-water (b)-LiBr(s) solution at 30°C

x_a	x_b	x_s	$\overline{V}_a$	$\overline{V}_b$	$\overline{V}_s$
0.4156	0.3168	0.2676	37.96	17.972	26.80
0.4002	0.1343	0.1859	39.04	18.102	23.69
0.3496	0.6218	0.0286	38.97	17.716	24.577
0.1892	0.7850	0.0258	38.11	17.996	25.894
0.0574	0.9192	0.0234	37.46	18.086	26.230

* *The partial molar volumes were obtained from the density formulas based on the data from Raatschen (1985).*

9.2. Find the furter constant k_F^0 from the water(1)-ethanol(2)-KI (salt= s) vapor-liquid equilibrium data at a constant pressure = 758 torr. The clean (salt free) solution has a constant mole fraction for ethanol in water at $x_2^0 = 0.309$. Salt mole fraction is x_s. To what value of x_s is the Furter equation valid?

9.3. From the data of Furter (Table 9E.2), find the Setchenov constant, k_S. The Setchenov equation is valid to what salt concentration x_s?

9.4. Convert the mole fraction $x_s = 0.1187$ to other concentration scales: molality M (basis 1000 g of clean solution water+ethanol), and molarity c (basis 1 liter of solution water+ethanol+KI). Find the density data from a handbook.[65]

9.5. Calculate the ionic strength I for the data in the table.

Table 9E.2. Isobaric Vapor-Liquid Equilibrium Data for the Potassium Iodide (s)-Ethanol (2)-Water (1) System at P=758 torr. Mole fraction of ethanol (a constant value) $x_2^0 = 0.309$ in the clean solution (salt free water-ethanol mixture). The ethanol mole fraction in vapor is y_2

x_2	y_2	T°C	$\log_{10}(\alpha s/\alpha 0)$
0	0.5844	82.0	0.0005
0.0100	0.5983	82.1	0.0263
0.0113	0.6044	82.2	0.0373
0.0218	0.6158	82.2	0.0581
0.0275	0.6233	82.2	0.0719
0.0344	0.6316	82.3	0.0873
0.0438	0.6437	82.6	0.1100
0.0488	0.6508	82.7	0.1235
0.0608	0.6634	82.9	0.1479
0.0687	0.6732	82.9	0.1670
0.0768	0.6839	83.0	0.1885
0.0823	0.6882	83.2	0.1971
0.0881	0.6943	83.5	0.2094
0.0992	0.7013	83.6	0.2238
0.1033	0.7112	83.8	0.2446
0.1187 (saturated)	0.7213	84.0	0.2662

*From W.F. Furter[35] 1979.

Chapter 10

Ionic Distributions:
An Integral Equation Approach

10.1. Introduction

The heart of the molecular theory of ionic solutions is the integral equation formulation for the probability distributions of ions in solution. Molecular distribution theory is a probabilistic description of how molecules (i.e. particles, ions, colloids) distribute in space and time given their interaction energies. Ever since the early solution[104] of the mean spherical approach (MSA) for Coulomb interactions in 1970s, the field of integral equations (IE) has developed by leaps and bounds. There are now many "brands" of integral equations for this purpose. One of the most successful IEs currently applied to ionic solutions is the hypernetted-chain (HNC) integral equation. We have cited the mean spherical approach (MSA) results in Chapter 6. Here we shall discuss other types of integral equations that can provide improved results.

Integral equations yield the distributions of ions in space, as singlet probability (distribution of a single particle as a function of r, the position vector of the molecule) and pair probability (distribution of a pair of particles as a function of r_1 and r_2, positions of the two molecules in space). These distribution functions can be directly related to the thermodynamic properties of the solution.[59] Thus starting from the knowledge of the potential energy of interaction, one can obtain the spatial distributions of ion using the integral equations. Next, one uses the molecular thermodynamic formulas[56] to obtain the solution properties from these distribution functions. By employing statistical mechanical formulas, a bridge is established between the microscopic world of interacting molecules and the macroscopic world of thermodynamic properties. The behavior of the ionic species is understood on a more fundamental level.

10.2. The Ornstein-Zernike Integral Equations and their Closures

We encourage the reader to review Chapter 6 on the Hamiltonian and the N-body probability density. Given the N-body probability, we can formulate the so-called *partition function* Z_N and the *pair density* $\rho^{(2)}(r_1, r_2)$ (the density of the probability of a pair of molecules situated at two locations r_1 and r_2 simultaneously).

$$Z_N \equiv \frac{1}{N!\Lambda^{3N}} \int dr^N \, \exp\left(-\beta V_N(r^N)\right) \equiv \frac{Q_N}{N!\Lambda^{3N}} \qquad (10.2.1)$$

where V_N is the total potential energy of the N-body system, and Λ is the *de Broglie thermal wave length*

$$\Lambda^2 \equiv \frac{h^2}{2\pi mkT} \qquad (10.2.2)$$

where h = Planck's constant (=6.62608 E-34 J s), m = molecular mass, k = Boltzmann constant (=1.38066 E-23 J/K). Q_N is the *configurational integral*

$$Q_N \equiv \int dr^N \, \exp\left[-\beta V_N(r^N)\right] \qquad (10.2.3)$$

The pair density is the *marginal probability* of the N-body probability when the N-2 position coordinates are integrated out leaving two chosen molecules (any two) occupying r_1, and r_2

$$\rho^{(2)}(\vec{r}_1, \vec{r}_2) \equiv \frac{N(N-1)}{Q_N} \int d\vec{r}_3 d\vec{r}_4 ... d\vec{r}_N \, \exp\left(-\beta V_N(r^N)\right) \qquad (10.2.4)$$

$\rho^{(2)}(r_1, r_2)$ denotes the probability (density) that gives *any* molecule found at the position , r_1, while there is another (any other) molecule found at r_2. We already know that the probability of finding a single molecule at , r_1 in a uniform isotropic fluid is simply the bulk number density (number of molecule per volume of the container) $\rho = N/V$; and that for a pair of non-correlated (non-interacting) molecules at , r_1 and , r_2 is ρ^2. We can "normalize" the pair density by ρ^2 to get the *pair correlation function (pcf)* or *radial distribution function (rdf)* $g^{(2)}(r_1, r_2)$

$$g^{(2)}(\vec{r_1},\vec{r_2}) \equiv \frac{\rho^{(2)}(\vec{r_1},\vec{r_2})}{\rho^2} = \frac{N(N-1)}{\rho^2 Q_N} \int d\vec{r_3}d\vec{r_4}...d\vec{r_N} \exp\left(-\beta V_N(r^N)\right) \quad (10.2.5)$$

Note that $g^{(2)}(r_1,r_2)$ is dimensionless. From $g^{(2)}(r_1,r_2)$ we can defined a number of other correlation functions that are useful in the IEs. Notably, the total correlation function $h(r_1,r_2)$, the cavity distribution function $y(r_1,r_2)$, the direct correlation function $c(r_1,r_2)$, and the indirect correlation function $\gamma(r_1,r_2)$.

$$h(\vec{r_1},\vec{r_2}) \equiv g^{(2)}(\vec{r_1},\vec{r_2}) - 1,$$
$$y(\vec{r_1},\vec{r_2}) \equiv g^{(2)}(\vec{r_1},\vec{r_2})\exp(+\beta u(\vec{r_1},\vec{r_2})), \quad (10.2.6a,b)$$

The Ornstein-Zernike (OZ) equation relates the direct correlation function $c(r_1,r_2)$ to the total correlation function $h(r_1,r_2)$:

$$h(\vec{r_1},\vec{r_2}) - c(\vec{r_1},\vec{r_2}) \equiv \int d\vec{r_3}h(\vec{r_1},\vec{r_3})\rho c(\vec{r_3},\vec{r_2}) \quad (10.2.7)$$

This is a convolution integral. The Ornstein-Zernike equation can be considered, given the total correlation $h(r_1,r_2)$, as a definition for the direct correlation $c(r_1,r_2)$. A deeper understanding of the OZ relation should be based on the reciprocal relation of the compressibility derivative and the number fluctuation derivative.[59] The indirect correlation $\gamma(r_1,r_2)$ is defined as

$$\gamma(\vec{r_1},\vec{r_2}) \equiv h(\vec{r_1},\vec{r_2}) - c(\vec{r_1},\vec{r_2}) \quad (10.2.8)$$

The OZ equation can be used to obtain the direct correlation $c(r_1,r_2)$. To do this, we need a *closure relation*, an independent second equation that connects once more the total correlation $h(r_1,r_2)$ to the direct correlation $c(r_1,r_2)$ (i.e., two equations solved for two unknown functions $h(r_1,r_2)$ and $c(r_1,r_2)$). The word "closure" means that it closes (or completes) the conditions for unique determination of $h(r_1,r_2)$ and $c(r_1,r_2)$. We use a very simple example to explain this connection. Consider the linear algebraic equation with two variables, x and y:

$$2x+3y =6$$

(x corresponds to $h(r_1,r_2)$, and y to $c(r_1,r_2)$). Clearly this equation (being considered as equivalent to the OZ situation) alone does not uniquely determine x and y. We need a second equation (i.e., an equivalent of the closure relation) relating x and y again: say,

$$ax+by = c \qquad \text{or} \qquad y=f(x)$$

This *closure equation* provides an additional condition for unique determination of the variables x and y!

From statistical mechanics (in the form of cluster expansions[59]), both $h(r_1,r_2)$ and $c(r_1,r_2)$ are expressed in terms of the Mayer bonds (bonds of the type $f(r) \equiv \exp[-\beta u(r)]-1$). Thus there is a *"functional"* (F) relation between $h(r_1,r_2)$ and $c(r_1,r_2)$. This functional can be used as the second relation: the closure equation. In mathematics, a functional, $h=F[c]$, with square brackets, is a relation where the value $h(r_1,r_2)$ on the left-hand side depends on all the domain values (r_1,r_2) of the function $c(r_1,r_2)$. Or the value of F, i.e. h, depends on the entire RXR domain where c is defined.

$$h(\vec{r}_1,\vec{r}_2) \equiv F[c(\vec{r}_1,\vec{r}_2)] \qquad (10.2.9)$$

One example of a functional is the area A of a function $f(x)$ situated between the limits a and b. A is a functional of f: $A= \varphi[f(x)]$. (A depends on all values of f within $a<x<b$, not just one value, say $f(3)$).

$$A \equiv \varphi[f(x)]= \int_a^b dx\, f(x)$$

Alternatively, we can rewrite the above functional F (10.2.9) with an equivalent functional G between the cavity function $y(r_1,r_2)$ and the indirect correlation $\gamma(r_1,r_2)$ (since both $y(r_1,r_2)$ and $\gamma(r_1,r_2)$ are related to h and c).

$$\ln y(\vec{r}_1,\vec{r}_2) \equiv G[\gamma(\vec{r}_1,\vec{r}_2)] \qquad (10.2.10)$$

From the theory of cluster series[59] (or alternatively from the functional expansion[60] of the singlet direct correlation $c^{(1)}(r_0)$, see below), this closure relation is often expressed[59] in terms of a bridge functional $B(r_1,r_2)$, i.e.

$$\ln y(\vec{r}_1, \vec{r}_2) \equiv G[\gamma(\vec{r}_1, \vec{r}_2)] = \gamma(\vec{r}_1, \vec{r}_2) + B[\gamma(\vec{r}_1, \vec{r}_2)] \qquad (10.2.11)$$

The bridge functional (as *defined* by (10.2.11)) has an exact "diagrammatical expansion" (a series in Mayer cluster diagrams). But this series is infinite and its high-order terms are extremely difficult to evaluate (e.g. for water, the first term $\rho^2 B_2$ is an eleven-dimensional integral). Mathematically, $B(r_1, r_2)$ is a "functional" of the correlation function γ. $B = B[\gamma(r_1, r_2)]$. B depends on the values of $\gamma(r_1, r_2)$ on RXR, not just on a single value of γ. Since 1950s, the usual practice in the liquid state theory was to simplify the relation by making a *unique function approximation*: that B is a function of γ. $B = B(\gamma)$, instead of a functional! As a consequence, many functions were proposed based on plausible arguments. The important ones that are commonly used in liquid state theory[59] are the Percus-Yevick (*PY*) approximate closure, and the hypernetted chain (*HNC*) approximate closure

$$
\begin{aligned}
B(\vec{r}_1, \vec{r}_2) &\cong \ln[1 + \gamma(\vec{r}_1, \vec{r}_2)] - \gamma(\vec{r}_1, \vec{r}_2), & (PY) \\
B(\vec{r}_1, \vec{r}_2) &\cong 0, & (HNC)
\end{aligned}
\qquad (10.2.12)
$$

A number of other approximate closures have also been proposed. The success or the lack thereof depends on the particular pair interaction of the molecular species. For example, for highly repulsive short-range interactions (e.g. hard spheres), the PY closure is highly accurate. But PY is poor for ionic species. Ionic interactions are long-ranged (Coulomb forces). HNC performs much better for the long-ranged forces. There is no general rule on the "goodness" of the closures. Since most existing closures are approximate, one has to examine their performance on a case by case basis.

10.3. The Numerical Solution Methods

Once we have the Ornstein-Zernike equation and its concomitant closure relation, we can devise numerical techniques to solve for the correlation functions $h(r_1, r_2)$ and $c(r_1, r_2)$. We shall discuss (i) the successive iteration method in real space (the Picard method); and (ii) the successive iteration method in Fourier space. There are other methods such as the Newton-Raphson method[55] with the Fourier series that are efficient and useful. We refer to them in the references.[55]

10.3.1. Successive substitutions – Picard's method

First, the OZ convolution integral can be written in the *bipolar coordinates* for isotropic fluids (whose interaction potentials are functions of r only and not of θ and ϕ) as

$$h(r) - c(r) \equiv \frac{2\pi\rho}{r} \int_0^\infty ds \ sc(s) \left[\int_{|r-s|}^{r+s} dt \ th(t) \right] \qquad (10.3.1)$$

where $r = |\mathbf{r}_2 - \mathbf{r}_1|$, $s = |\mathbf{r}_3 - \mathbf{r}_1|$, and $t = |\mathbf{r}_3 - \mathbf{r}_2|$. This double integral can be evaluated by the following steps. First, the inner integral is calculated via the $E(x)$ function defined by

$$E(x) \equiv \int_0^x dt \ th(t) \qquad (10.3.2)$$

Thus the convolution integral (10.3.1) can be written as

$$h(r) - c(r) \equiv \frac{2\pi\rho}{r} \int_0^\infty ds \ sc(s) \ [E(r+s) - E(|r-s|)] \qquad (10.3.3)$$

We next use the closure relation

$$\ln y(r) = \ln[(1 + h(r))\exp(\beta u(r))] = \gamma(r) + B(\gamma(r)) = h(r) - c(r) + B((h-c)) \qquad (10.3.4)$$

If we choose the Percus-Yevick closure, we have

$$\ln y(r) = \ln[(1 + h(r))\exp(\beta u(r))] \cong \ln[1 + h(r) - c(r)] \qquad (PY) \ (10.3.5)$$

We observe that the pair potential $u(r)$ appears in the closure (10.3.5). The temperature ($\beta = 1/(kT)$ and density ρ of the system appear in the closure and the OZ, respectively.

The numerical procedure starts, as is usually the case, with an initial guess of the total and direct correlations (we do not know their precise values, so we make an intelligent guess).

$$h^0(r) \text{ and } c^0(r) \qquad (10.3.6)$$

At low densities, the cluster theory says that

$$h^0(r) \cong Mayer \ f(r) = \exp[-\beta u(r)] - 1, \quad and \quad c^0(r) \cong -\beta u(r) \qquad (10.3.7)$$

These approximations can be used as the initial guess functions $h^0(r)$ and $c^0(r)$ at low densities (i.e. at $\rho\sigma^3$ <0.01). For higher densities, on can build the solutions $h(r)$ and $c(r)$ from low densities ($\rho\sigma^3$ from 0.01, 0.02, 0.03, ... , 0.1, 0.15, 0.20, ...) gradually up to higher densities. Use the solutions at present density as inputs for the next higher density (e.g. from $\rho\sigma^3 = 0.09$ to 0.1, or from 0.20 to 0.21).

The numerical solution procedure is outlined as: (i) Substitute the $h^0(r)$ and $c^0(r)$ to the right-hand side of OZ (10.3.3) to get a new $h^1(r)$. (ii) This new $h^1(r)$ is substituted into the right-hand side of the closure (10.3.9) to get a new $c^1(r)$. The OZ and closure equations can be rearranged to give

$$h^1(r) = c^0(r) + \frac{2\pi\rho}{r} \int_0^\infty ds \ sc^0(s) \ \left[E^0(r+s) - E^0(|r-s|)\right] \qquad (OZ) \quad (10.3.8)$$

$$c^1(r) = \left[1 + h^1(r) - (1 + h^1(r))\exp(\beta u(r))\right] \qquad (PY) \quad (10.3.9)$$

(iii) These new $h^1(r)$ *and* $c^1(r)$ are substituted to the right-hand side of the OZ (10.3.8) and the closure (10.3.9)) again to obtain the new $h^2(r)$ and $c^2(r)$ (the second outputs). The superscript 2 means 2nd, not "squared". Next, from $h^2(r)$ and $c^2(r)$, we get $h^3(r)$ and $c^3(r)$. This method of successive substitutions is called the *Picard method*. It is repeated until a prescribed convergence criterion is satisfied. According to calculus, the Cauchy condition for absolute convergence of infinite sequence of functions is that the successive absolute difference between the *n*th iteration and the *(n+1)*th iteration is less than a prescribed small positive number ε

$$|\gamma^{n+1}(r) - \gamma^n(r)| < \varepsilon, \quad \forall r \qquad (10.3.10)$$

If ε is very small, the condition of convergence is very stringent. Many iterations will be needed (sometimes over 1000 iterations). If ε is large, then the sequence may not have converged completely. (Namely we do not obtain the correct final answer). The results will contain substantial errors. For practical purposes, $\varepsilon \sim 0.001$ is quite sufficient for convergence. Note first of all, we have selected in (10.3.10) the indirect correlation $\gamma(r)$ as the test function, because $\gamma(r)$ gives a more stringent test than either h or c. Second, the absolute values of differences are tested at all grid points of r ($0<r< r_{max}$, $r_{max} = 6\sigma$ or 20σ). Normally, we discretize the r-value into grids, spanning the range of the interaction potential. For Lennard-Jones potential, the range L of significant (non-zero) interactions is about 6σ ($L= 6$ times the diameter). Thus for $0< $ r $<6\sigma$, we cut the range into 600 grids with grid size $\Delta r = 0.01\sigma$. We have N=601 grid points at r/σ =0.00, 0.01, 0.02, ... 1.00, 1.01, ..., 5.99, 6.00. In computer programming, both the distance r and the function $\gamma_i= \gamma(r_i)$, $i=0,1,2, ..., 600$ are discretized vectors (arrays). For example:

Dimension r(0:600), gamma(0:600)
r(i) = 0.00, 0.01, 0.02, ..., 5.99, 6.00
gamma(i) = 20.3, 19.2, 18.6, ... , -0.001, 0.003, 0.000

When testing the convergence, each and every γ_i must satisfy the convergence criterion according to eq.(10.3.10) (i.e., $|\gamma_i^{n+1}- \gamma_i^{n}| < \varepsilon$, for all $i=0,...,600$). The number of iterations, n, (how many times we carry out the Picard substitutions) must reach a large number in order to achieve convergence (n is normally 50~100, and at times 300~500). In principle, large L and small Δr ensure numerical accuracy. However, these stringent choices are very time-consuming. Only when the density is high and temperature is low, we need to increase the range of integration L (to 10σ or more). The pair correlation function $g(r)$ grows in magnitude and range at low temperatures and high densities. To obtain an optimum combination of L and Δr, some trial-and-error is involved. The determination of convergence is based on the γ-function, or the virial pressure value, that they do not change appreciably (say, less thane 1%) when L and Δr are tightened. For ionic solutions, L should be even larger, due to the long-range Coulomb potential. At least $L =100\ \sigma$ or $200\ \sigma$ is needed. If the pair potential varies slowly with r, a large Δr is tolerable. However, for fast varying pair potentials (such as in hydrogen bonding), small Δr is required to capture the rapid changes in the energy.

We note that in the above outline, it is immaterial which functions $c(r)$ or $h(r)$ is solved first. We have in eqs.(10.3.8 & 9) solved for $h(r)$ first, and used the closure to find $c(r)$. We could have equally well solved for $c(r)$ first, then used the closure to find $h(r)$. The answers in principle should be the same. For many cases the *direct* successive substitutions as outlined above lead to numerical instability (the iterations produce increasingly large oscillating γ_i values, eventually reaching infinity—i.e. divergence), while a true solution may actually exist for the given temperature and density. We suggest a mixing (relaxation) scheme to assure numerical stability: i.e., we mix the input functions with the output functions before the next substitution into OZ. This way we reduce the large "shocks" arising from a vastly different input.

First inputs = $h^0(r)$ and $c^0(r)$,
  Go to OZ and closure →
  Outputs = $h^1(r)$ and $c^1(r)$

Second inputs: $h^{in2}(r) = \alpha\, h^0(r) + (1\text{-}\alpha)\, h^1(r)$ **(mixing)**   (10.3.11)
          $c^{in2}(r) = \alpha\, c^0(r) + (1\text{-}\alpha)\, c^1(r)$ **(mixing)**
  Go to OZ and closure →
  Outputs: $h^2(r)$ and $c^2(r)$

Third inputs: $h^{in3}(r) = \alpha\, h^{in2}(r) + (1\text{-}\alpha)\, h^2(r)$ **(mixing)**
          $c^{in3}(r) = \alpha\, c^{in2}(r) + (1\text{-}\alpha)\, c^2(r)$ **(mixing)**
  Go to OZ and closure →
  Outputs: $h^3(r)$ and $c^3(r)$

(This procedure is continued until convergence). Note that α is a mixing parameter with values between 0 and 1.

$$0 < \alpha < 1 \qquad (10.3.12)$$

Thus if $\alpha = 0$, the procedure becomes the direct substitution (as in Picard). If $\alpha = 1$, then there is no progress. By choosing $0 < \alpha < 1$, one can achieve better numerical stability and for most cases assure convergence to a final γ. The α value we normally use is about 0.5 to 0.85. In more difficult cases (e.g., $T^* = kT/\varepsilon \sim 0.73$, and $\rho^* = \rho\sigma^3 \sim 0.80$—for Lennard-Jones potential), using an α-value of 0.99 is not unusual with the mixing method. It all depends on how "close" the input functions are to the "final" converged solution.

10.3.2. Successive substitutions – in Fourier space

For convolution integrals, the Fourier transformation possesses the desirable property of changing a convolution into a simple product of the transformed functions. The OZ equation can be solved in the Fourier k-space by simple algebra. Applying Fourier transformation to eq.(10.2.7)

$$\tilde{h}(k) - \tilde{c}(k) = \tilde{h}(k)\rho\tilde{c}(k) \qquad (10.3.13)$$

where the tilde ~ represents the three-dimensionally Fourier-transformed function, and k is the reciprocal vector in the Fourier space conjugate to r in the real space. To be precise

$$\tilde{h}(k) \equiv \frac{4\pi}{k}\int_0^\infty dr\, rh(r)\sin(kr) \qquad (10.3.14)$$

Eq.(10.3.13) can be solved for $\tilde{h}(k)$

$$\tilde{h}(k) = \frac{\tilde{c}(k)}{1-\rho\tilde{c}(k)}, \qquad OZ\ in\ Fourier\ space \qquad (10.3.15)$$

This is the OZ equation in Fourier space. Thus given $c^0(r)$ in real space, we take the Fourier transform $\tilde{c}^0(k)$ and use eq.(10.3.15) to obtain $\tilde{h}^1(k)$. To apply the closure relation, we need to transform back to the real space, because the closure equation is usually nonlinear (it cannot be transformed into the Fourier space). The three-dimensional inverse Fourier transform of any function is

$$h(r) \equiv \frac{1}{2\pi^2 r}\int_0^\infty dk\, k\tilde{h}(k)\sin(kr) \qquad (10.3.16)$$

Nowadays we have at our disposal the *Fast Fourier Transform (FFT)*, a very efficient forward and backward Fourier transform computer code. We can efficiently transform $\tilde{h}^1(k)$ back to $h^1(r)$ by FFT. Next we apply the closure relation (10.3.9) (for PY) in real space to get $c^1(r)$. This completes the first Picard iteration! Next we apply the mixing method of (mixing $\tilde{c}^0(k)$ with $\tilde{c}^1(k)$ by a parameter α as in eq.(10.3.11)). The process is repeated (via (10.3.15), (10.3.16), (10.3.9), and (10.3.11)) for as many iterations as are needed until convergence is attained. For Coulomb potential, the interaction is long-ranged (reaching 100σ to 200σ, σ being the cation or anion diameter). The number of grids in Fast Fourier Transform should be a power of 2. Use 8192 or 16384 grids with grid size $\Delta r \sim 0.01$ σ.

10.3.3. Renormalization/Optimization of the direct correlation c(r)

There are numerical problems associated with the Fourier transform of the long-ranged $h(r)$ and $c(r)$ ion-ion correlation functions. The integrals may not converge (the calculated values of the integrals continue to grow upon increasing the range of integration). For ionic interactions, one can compensate by cleverly incorporating the Debye-Hückel type potentials into the total correlation $h(r)$ and the direct correlation $c(r)$. This procedure is called "renormalization" or "optimization". The direct correlation $c(r)$ is divided into a short-range part $c^{SR}(r)$ and a long-range part $c^{LR}(r)$.

$$c(r) \equiv c^{SR}(r) + c^{LR}(r) \tag{10.3.17}$$

We have omitted the species designations (cation or anion) to simplify writing. The long-range part is approximated by the Coulomb potential,

$$c_{ij}^{LR}(r) \cong -\beta u_{ij}^{Coulomb}(r) = -\frac{q_i q_j}{\varepsilon kTr} = -\frac{z_i z_j e^2}{\varepsilon kTr} \tag{10.3.18}$$

Whatever is left over after subtraction of the Coulomb potential from $c(r)$ is called the short-range direct correlation $c^{SR}(r)$. The total correlation is also separated into two parts: one long-ranged and the other short-ranged

$$h(r) \equiv h^{SR}(r) + h^{LR}(r) \tag{10.3.19}$$

The long-range part is chosen to be the Debye screened potential

$$h_{ij}{}^{LR}(r) \cong \exp\left[-\frac{q_i q_j}{\varepsilon kTr} e^{-\kappa r}\right] - 1 = \exp\left[-\frac{z_i z_j e^2}{\varepsilon kTr} e^{-\kappa r}\right] - 1, \qquad (10.3.20a)$$

At large r the exponents can be expanded and give

$$h_{ij}{}^{LR}(r) \cong \left[-\frac{q_i q_j}{\varepsilon kTr} e^{-\kappa r}\right] = \left[-\frac{z_i z_j e^2}{\varepsilon kTr} e^{-\kappa r}\right] \qquad (10.3.20b)$$

where κ is the inverse shielding length defined in Chapter 4. There are other alternative ways in literature to achieve this renormalization/optimization (separations into the long-range h^{LR} and c^{LR}). As a matter of terminology, the division of the dcf $c(r)$ into c^{SR} + c^{LR} is called the "*optimization*", while the division of the tcf $h(r)$ into h^{SR} + h^{LR} is called "*renormalization*". We have used here a simple procedure (10.3.18) and (10.3.20) that functions well in avoiding the divergence in Fourier transformation. We cite one example of optimization due to Duh et al.[26] They chose the long-range part $c^{LR}(r)$ as

$$c_{ij}{}^{LR}(r) \cong \left[1 - e^{\beta u_{ij}^{SR}(r)}\right] h_{ij}^{SR}(r) - \frac{q_i q_j}{\varepsilon kTr} \qquad (10.3.21)$$

This sophisticated renormalization procedure was considered to be "better behaved". (the $u^{SR}(r)$ above was the short-range part of the pair potential.) The short-range $h^{SR}(r)$ was obtained from a pseudo-Ornstein-Zernike equation connecting $h^{SR}(r)$ to $c^{SR}(r)$.

The Fourier transformation of $c(r)$ and $h(r)$ is carried out by first transforming the long-range parts. The Fourier transforms of the long-range Coulomb potential and the Debye screened potential are known.

For x(r) = (1/r),

$$\tilde{x}(k) \equiv \frac{4\pi}{k} \int_0^\infty dr \; rx(r) \sin(kr) = \frac{4\pi}{k^2} \qquad (10.3.22)$$

For λ(r) = [exp(-κr)/r],

$$\tilde{\lambda}(k) \equiv \frac{4\pi}{k} \int_0^\infty dr\, r\lambda(r)\sin(kr) = \frac{4\pi}{\kappa^2 + k^2} \qquad (10.3.23)$$

To transform $c(r)$, we write

$$c_{ij}(r) \equiv c_{ij}^{SR}(r) + c_{ij}^{LR}(r) = c_{ij}^{SR}(r) - \frac{q_i q_j}{\varepsilon_m kTr} \qquad (10.3.24)$$

We transform the two functions $c^{SR}(r)$ and $c^{LR}(r)$, separately

$$c_{ij}^{SR}(r) \equiv c_{ij}(r) + \frac{q_i q_j}{\varepsilon kTr} \quad \rightarrow \text{Fourrier Transformed to} \quad \tilde{c}_{ij}^{SR}(k)$$

$$c_{ij}^{LR}(r) \equiv -\frac{q_i q_j}{\varepsilon kTr} \quad \rightarrow \text{Fourrier Transformed to} \quad \tilde{c}_{ij}^{LR}(k) = -\frac{4\pi z_i z_j e^2}{\varepsilon_m kTk^2} \qquad (10.3.25a,b)$$

It is possible now to transform $c^{SR}(r)$ without divergence, since after adding the Coulomb term to $c(r)$, $c^{SR}(r)$ is short-ranged. By adding the two transformed pieces together, we obtain the full $\tilde{c}_{ij}(k)$ in Fourier-space

$$\tilde{c}_{ij}(k) = \tilde{c}_{ij}^{SR}(k) + \tilde{c}_{ij}^{LR}(k) \qquad (10.3.26)$$

Similar treatment is carried out for $\tilde{h}_{ij}(k)$

$$h_{ij}^{SR}(r) \equiv h_{ij}(r) + \frac{q_i q_j}{\varepsilon kTr}e^{-\kappa r} \quad \rightarrow \text{Fourrier Transformed to} \quad \tilde{h}_{ij}^{SR}(k)$$

$$h_{ij}^{LR}(r) \equiv -\frac{q_i q_j}{\varepsilon kTr}e^{-\kappa r} \quad \rightarrow \text{Fourrier Transformed to} \quad \tilde{h}_{ij}^{LR}(k) = -\frac{4\pi z_i z_j e^2}{\varepsilon kT(\kappa^2 + k^2)}$$

$$(10.3.27)$$

$$\tilde{h}_{ij}(k) = \tilde{h}_{ij}^{SR}(k) + \tilde{h}_{ij}^{LR}(k) \qquad (10.3.28)$$

10.4. The Hypernetted Chain Closure

Since the hypernetted-chain closure[101] gives quite accurate answers for ionic solutions (yielding reasonable electrostatic energies, activity coefficients and generally good structures $g(r)$), we discuss its performance here. There is a rich literature on the integral equation studies of the ionic fluids. Carley[16] in 1967 solved the HNC and PY formulations for the primitive electrolytes. Since then, Friedman,[30] Rasaiah,[86] Rossky,[91] Allnatt,[1] Henderson,[95] Hafskjold,[41] Valleau,[101] and their coworkers have formulated various improvements and methods of solution to the OZ equations. The HNC closure sets $B=0$. Thus

$$\ln y_{ij}(r) = \ln\{[1 + h_{ij}(r)]\exp[\beta u_{ij}(r)]\} \cong h_{ij}(r) - c_{ij}(r), \quad i,j = +,- \qquad (10.4.1)$$

The OZ equation can be solved with this closure by the methods outlined above. For the primitive model (PM) of electrolyte solutions, the ions are considered as charged hard spheres, with the solvent molecules removed and supplanted in their place by a "dielectric continuum" of the same permittivity ε_m. The three interaction potentials are

Cation-cation potential with cation-cation collision diameter σ_{++}

$$u_{++}(r) = \frac{z_+ z_+ e^2}{\varepsilon_m r}, \qquad if \quad r > \sigma_{++}$$
$$u_{++}(r) = \infty, \qquad if \quad r \le \sigma_{++} \qquad (10.4.2)$$

Cation-anion potential with cation-anion collision diameter σ_{+-}

$$u_{+-}(r) = \frac{z_+ z_- e^2}{\varepsilon_m r}, \qquad if \quad r > \sigma_{+-}$$
$$u_{+-}(r) = \infty, \qquad if \quad r \le \sigma_{+-} \qquad (10.4.3)$$

Anion-anion potential with anion-anion collision diameter σ_{--}

$$u_{--}(r) = \frac{z_- z_- e^2}{\varepsilon_m r}, \qquad if \quad r > \sigma_{--}$$
$$u_{--}(r) = \infty, \qquad if \quad r \le \sigma_{--} \qquad (10.4.4)$$

In general, the diameters σ_{++}, σ_{+-}, and σ_{--} are all different. For example, the cation Na$^+$ may have a diameter σ_{++} = 1.90 Å, and the anion Cl$^-$ σ_{--}=3.62 Å. The cation-anion diameter on average is σ_{+-}= 2.76 Å.

There are three distinct OZ equations in Fourier space, one for each for the cation-cation (11), cation-anion (12), and anion-anion (22) pairs

$$\tilde{h}_{11}(k) - \tilde{c}_{11}(k) = \tilde{h}_{11}(k)\rho_1 \tilde{c}_{11}(k) + \tilde{h}_{12}(k)\rho_2 \tilde{c}_{21}(k) \qquad (10.4.5)$$

$$\tilde{h}_{12}(k) - \tilde{c}_{12}(k) = \tilde{h}_{11}(k)\rho_1 \tilde{c}_{12}(k) + \tilde{h}_{12}(k)\rho_2 \tilde{c}_{22}(k) \qquad (10.4.6)$$

$$\tilde{h}_{22}(k) - \tilde{c}_{22}(k) = \tilde{h}_{21}(k)\rho_1 \tilde{c}_{12}(k) + \tilde{h}_{22}(k)\rho_2 \tilde{c}_{22}(k) \qquad (10.4.7)$$

There is also the "21" pair in OZ (which is equivalent to the 12-pair eq.(10.4.6)).

$$\tilde{h}_{21}(k) - \tilde{c}_{21}(k) = \tilde{h}_{21}(k)\rho_1 \tilde{c}_{11}(k) + \tilde{h}_{22}(k)\rho_2 \tilde{c}_{21}(k) \qquad (10.4.8)$$

We need only one of the two equations (the pair 12 or the pair 21) for solution, because they are equivalent to each other by reason of symmetry (12 ⇔ 21). Choosing any one of them is sufficient.

The above four equations can be written in a matrix form

$$\begin{bmatrix} \tilde{h}_{11} & \tilde{h}_{12} \\ \tilde{h}_{21} & \tilde{h}_{22} \end{bmatrix} - \begin{bmatrix} \tilde{c}_{11} & \tilde{c}_{12} \\ \tilde{c}_{21} & \tilde{c}_{22} \end{bmatrix} = \begin{bmatrix} \tilde{h}_{11} & \tilde{h}_{12} \\ \tilde{h}_{21} & \tilde{h}_{22} \end{bmatrix} \begin{bmatrix} \rho_1 & 0 \\ 0 & \rho_2 \end{bmatrix} \begin{bmatrix} \tilde{c}_{11} & \tilde{c}_{12} \\ \tilde{c}_{21} & \tilde{c}_{22} \end{bmatrix} \qquad (10.4.9)$$

Let $\underline{\underline{h}}$, $\underline{\underline{c}}$, and $\underline{\underline{\rho}}$ be the matrices

$$\underline{\underline{h}} \equiv \begin{bmatrix} \tilde{h}_{11} & \tilde{h}_{12} \\ \tilde{h}_{21} & \tilde{h}_{22} \end{bmatrix}, \qquad \underline{\underline{c}} \equiv \begin{bmatrix} \tilde{c}_{11} & \tilde{c}_{12} \\ \tilde{c}_{21} & \tilde{c}_{22} \end{bmatrix}, \qquad \underline{\underline{\rho}} \equiv \begin{bmatrix} \rho_1 & 0 \\ 0 & \rho_2 \end{bmatrix} \qquad (10.4.10)$$

The OZ equations can be written in matrix notation

$$\underline{\underline{h}} - \underline{\underline{c}} = \underline{\underline{h}}\,\underline{\underline{\rho}}\,\underline{\underline{c}} \qquad (10.4.11)$$

We can simplify the primitive model to a *restricted* primitive model (*RPM*) by requiring that $\sigma_{++} = \sigma_{+-} = \sigma_{--} = \sigma$. Namely, all ions have the same diameter σ. Then only two of the three OZ equations are needed: (10.4.5) and (10.4.6).

$$\tilde{h}_{11}(k) - \tilde{c}_{11}(k) = \tilde{h}_{11}(k)\rho_1 \tilde{c}_{11}(k) + \tilde{h}_{12}(k)\rho_2 \tilde{c}_{21}(k) \qquad (10.4.5)$$

$$\tilde{h}_{12}(k) - \tilde{c}_{12}(k) = \tilde{h}_{11}(k)\rho_1 \tilde{c}_{12}(k) + \tilde{h}_{12}(k)\rho_2 \tilde{c}_{22}(k) \qquad (10.4.6)$$

The anion-anion pair is now equivalent to the cation-cation pair, $h_{22}(r) = h_{11}(r)$ (in RPM).

Example of an RPM electrolyte

In literature there have been two specific models of equal-sized ions (i) hard sphere repulsion + Coulomb potential and (ii) soft inverse-9th power repulsion + Coulomb potential.[5] The former is given by equations (10.4.2 & 10.4.3) with a hard sphere core chosen to be $\sigma_{++} = 4.2$ Å. The latter[91] is shown below. The repulsive part is an inverse-9th power term.

$$u_{ij}(r) = \frac{kb \mid z_i z_j \mid}{\sigma_{ij}}\left(\frac{\sigma_{ij}}{r}\right)^9 + \frac{z_i z_j e^2}{\varepsilon_m r}, \quad i, j = +,- \qquad (10.4.12)$$

where k is the Boltzmann constant, b is chosen to be 5377.75 (Å·Kelvin), $\sigma = 2.8428$ Å. At 25°C the relative dielectric constant D= 78.358 (thus $\varepsilon_m = 78.358\varepsilon_0$, ε_0 = the permittivity of vacuum). The soft inverse 9th power potential is the short-range $u_{ij}^{SR}(r)$. With these parameters, the minimum value of $u_{+-}(r)$ of the soft RPM is also at $r_{min} = 4.2$ Å. It has been shown that with these parameters, the hard RPM and the soft RPM are similar to each other in terms of structure and thermodynamic properties. A graphical comparison[5] of the soft RPM potential (10.4.12) with the charged hard sphere potential is given in Figure 10.4.1. The concentration range is chosen from very dilute solutions 0.001M, 0.005M, 0.0625M, to 0.2M. The electrolyte is of the 2-2 type. The HNC equations were solved according to the procedures outlined above. We present the results[26] below.

Behavior of the structures $g_{ij}(r)$

Figures 10.4.2 to 10.4.5 show the like (++,– –) and unlike (+–) pair correlation functions $g_{ij}(r)$ as obtained from the HNC closure[26] at four concentrations: 0.001M, 0.005M, 0.0625M, to 0.2M. Together are plotted the Monte-Carlo (MC) or molecular dynamics (MD) results which act as standards for comparison. Other lines represent results from *PYA, IPYO, INV,* and *LM,* some alternative theories.[26] (INV is the results from the inversion of the MD data, and is thus equivalent to the MD results).

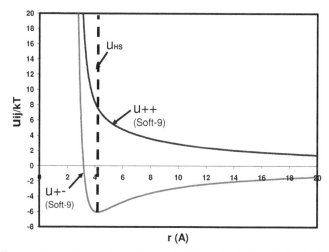

Figure 10.4.1. Comparison of the soft RPM potential (solid lines) with the hard RPM potential (dashed lines). d = 4.2 Å. They share the same potential minimum.[5]

In Figure 10.4.2, the solution is very dilute (0.001M). The HNC $g_{++}(r)$ (dashed lines) shows a prominent peak (a first maximum) at $r \sim 8$ Å (twice the r_{min} =4.2 Å). The location seems to indicate the possible formation of linear triplets (OOO, a chain of cation-anion-cation). The two cations are separated in-between by an anion (vice versa). However, the MD simulation gives no indication of such triplet state. It is recognized nowadays that this triplet state is spurious. HNC drastically overestimates the occurrence of triplet ions. As for the unlike pair correlation $g_{+-}(r)$, HNC underestimates the MD results. Thus HNC is not accurate at this very low concentration. Note that the value of the first peak of cation-anion g_{+-} (r) is very high, 234 (MD) vs. 186 (HNC).

At 0.005M (Figure 10.4.3), the discrepancy in HNC persists. Again HNC shows a pronounced first maximum in the like-pair $g_{++}(r)$ that is absent from the MD. For $g_{+-}(r)$, the peak of HNC (85) is much lower than the MD value (112).

At 0.0625M (Figure 10.4.4), the performance of HNC is improved. The first maximum of the like-pair $g_{++}(r)$ from HNC disappears. It is more in line with the MD data. However, the HNC curve underestimates the MD curve. For the unlike-pair g_{+-} (r), the peak of HNC (18.4) is close to (but again lower than) the MD value (22.5).

At 0.2M (Figure 10.4.5), HNC performs better. Thus HNC is a reasonable theory at high concentrations for electrolytes (for both 1-1 and 2-2 types). The g_{+-} (r) peak of HNC is 8.81, and MD 9.60.

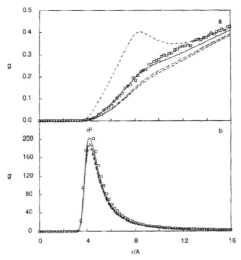

Figure 10.4.2. The like-pair (a) and unlike-pair (b) radial distribution functions $g_{++}(r)$ and $g_{+-}(r)$, respectively (1) From HNC closure (dashed lines), and (2) MD simulation (squares). Molality at 0.001M. (Other lines are from modified theories: PYA, IPYO, INV, and LM). (Duh et al.[26])

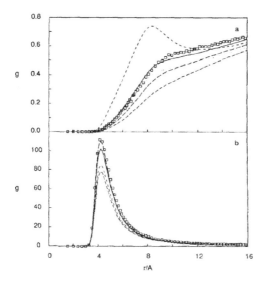

Figure 10.4.3. The like-pair (a) and unlike-pair (b) radial distribution functions $g_{++}(r)$ and $g_{+-}(r)$, respectively (1) From HNC closure (dashed lines), and (2) MD simulation (squares). Molality at 0.005M. (Duh et al.[26])

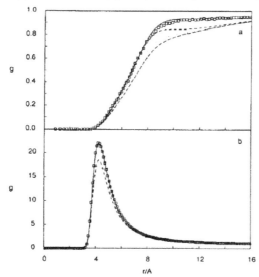

Figure 10.4.4. The like-pair (a) and unlike-pair (b) radial distribution functions $g_{++}(r)$ and $g_{+-}(r)$, respectively (1) From HNC closure (dashed lines), and (2) MD simulation (squares). Molality at 0.0625M. (Duh et al.[26])

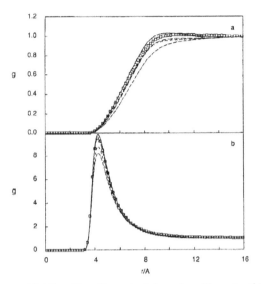

Figure 10.4.5. The like-pair (a) and unlike-pair (b) radial distribution functions $g_{++}(r)$ and $g_{+-}(r)$, respectively (1) From HNC closure (dashed lines), and (2) MD simulation (squares). Molality at 0.20M.. (Duh et al.[26])

Thermodynamic properties of HNC

We can calculate the electrostatic energy and the osmotic coefficient from the pair correlation function $g_{ij}(r)$ of HNC. Recall that the molecular-level formulas for the energy and the osmotic pressure are

Electrostatic internal energy, U^{ES}

$$\frac{U^{ES}}{VkT} = \frac{1}{2kT}\sum_{i,j}\rho_i\rho_j\int_0^\infty dr\, 4\pi r^2 u_{ij}(r)g_{ij}(r) \qquad (10.4.13)$$

Osmotic coefficient, ϕ

$$\phi = \frac{P^{osm}}{\rho_{TOT}kT} = 1 - \frac{1}{6\rho_{TOT}kT}\sum_{i,j}\rho_i\rho_j\int_0^\infty dr\, 4\pi r^3 g_{ij}(r)\frac{du_{ij}(r)}{dr} \qquad (10.4.14)$$

The internal energy calculated is given in Table 10.4.1. The osmotic coefficient is given in Table 10.4.2. For the electrostatic energy $U^{ES}/\rho_{TOT}kT$, the HNC values are smaller than the MD data at low concentrations (0.001M, 0.005M), but improve as the concentration increases (to 0.5625M).

Table 10.4.1. The electrostatic energy $-U^{ES}/\rho_{TOT}kT$. *(Duh.[26])*

c(M)	MD	HNC
0.001	0.469 ± 0.004	0.4328
0.005	0.9226	0.8285
	0.889 ± 0.045	
0.020	1.413	1.279
0.0625	1.834 ± 0.004	1.713
0.200	2.255 ± 0.003	2.197
	2.270 ± 0.024	[2.259]
0.5625	2.666 ± 0.016	[2.732]

Table 10.4.2. The osmotic coefficient $\phi = P^{osm}/\rho_{TOT}kT$ *(via virial eq.).* *(Duh.[26])*

c(M)	MD	HNC
0.001	0.898	0.890
0.005	0.808	0.802
0.020	0.706	0.713
0.0625	0.646 ± 0.003	0.642
0.200	0.611 ± 0.002	0.594

The agreement of the osmotic coefficient $\phi = P^{osm}/\rho_{TOT}kT$ derived from HNC is good from low to high concentrations (0.001M to 0.5625M). Overall, the HNC thermodynamic properties are quite dependable.

10.5. The Behavior of the Bridge Functions for Molten Salts

We have shown that the HNC closure is reasonable for electrolyte solutions at moderate to high concentrations (m > 0.1M). It fails dramatically for dilute solutions, especially for 2-2 electrolytes. This comes as a surprise, since most approximate closures are poor at high densities, but performs well at low densities. Furthermore, what is the situation for molten salts? (Salts at high temperatures will melt and produce ions.) There the McMillan-Mayer scale is the correct picture (as there is no solvent.) Recall that HNC assumes that the bridge function

$B_{ij}(r)$ is identically zero at any densities and temperatures! If we can show that $B_{ij}(r)$'s are not zero for some real systems, then HNC closure is not universal, and cannot be applied without discrimination.

One way to test this is to carry out molecular simulations (Monte-Carlo (MC) method or molecular dynamics (MD)) for a molten salt system and find the bridge functions that are capable of interpreting the structural data. These bridge functions will tell us whether the HNC hypothesis is of general validity. Tasseven et al.[100] have performed such a task. They carried out MD simulations for the molten salts of sodium chloride and silver iodide. They inverted the MD structures $g_{ij}(r)$ to obtain the bridge functions. We briefly describe their procedure below.

The pair potentials used for NaCl and AgI were of the Born-Huggins-Mayer type.

$$u_{ij}^{NaCl} = b_{ij} e^{c(\sigma_i + \sigma_j - r)} + \frac{z_i z_j e^2}{r} + \frac{C_{ij}}{r^6} - \frac{D_{ij}}{r^8} \qquad (10.5.1)$$

where $c = 3.115 \text{Å}^{-1}$, $\sigma_+ = 1.170 \text{Å}$, $\sigma_- = 1.585 \text{Å}$, $z_+ = 1$, $z_- = -1$. The parameters C_{ij}, D_{ij} and b_{ij}, are given in Table 10.5.1. A slightly different potential was used for AgI.

$$u_{ij}^{AgI} = \frac{H_{ij}}{r^{\eta_{ij}}} + \frac{z_i z_j e^2}{r} + \frac{P_{ij}}{r^4} + \frac{C_{ij}}{r^6} \qquad (10.5.2)$$

where $z_+ = 0.6$, $z_- = -0.6$. The parameters are given in Table 10.5.2 below.

Table 10.5.1 The Parameters in the Born-Huggins-Mayer Potential for NaCl

	b_{ij} (eV)	C_{ij} (eV Å^6)	D_{ij} (eV Å^8)
++	0.264	1.05	0.499
+−	0.211	6.99	8.68
− −	0.158	72.4	145.5

Table 10.5.2 The Parameters in the Born-Huggins-Mayer Potential for AgI

	η_{ij}	H_{ij} (eV Å$^{\eta_{ij}}$)	P_{ij} (eV Å^4)	C_{ij} (eV Å^6)
++	11	0.2132	0	0
+−	9	1548.5	16.9	0
−−	7	6431.5	33.8	99.8

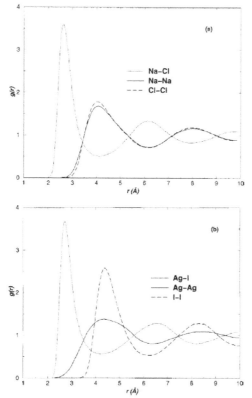

Figure 10.5.1. The radial distribution functions of the molten salts (a) NaCl and (b) AgI, obtained from molecular dynamics simulations.[100]

The temperature for the molten NaCl was chosen to be T = 1165K, and density ρ = 0.0314 ions/Å^3. For AgI, T = 933K, and ρ = 0.0281 ions/Å^3. These values are near their melting points. For molten salts, without solvents, the relative dielectric constant is D=1 (with the permittivity ε_0 of vacuum). In the MD simulation, N= 216 ions (with 108 cations and 108 anions) were placed in a cubic box with periodic boundary condition. The time step was 5 fs. Ewald sum was used to account for the long-range Coulomb interactions. The pair correlation functions $g_{ij}(r)$ were obtained. They are shown in Figure 10.5.1. We see that the Na$^+$-Cl$^-$ pair $g_{+-}(r)$ has a first peak at a short distance r~ 2.8Å; so does the Ag$^+$-I$^-$ $g_{+-}(r)$ peak. The I$^-$-I$^-$ $g_{--}(r)$ peak is at r~ 4.5 Å and is much higher than the Ag$^+$-Ag$^+$ $g_{++}(r)$ peak.

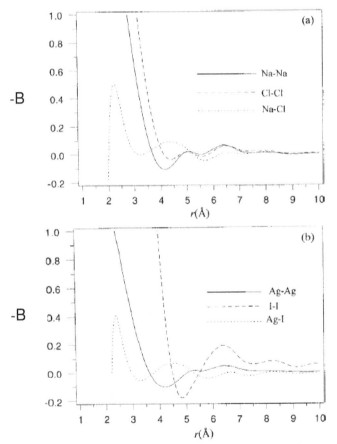

Figure 10.5.2. The bridge functions −B(r) (minus B) of the molten salts (a) NaCl, and (b) AgI, obtained from inverting the molecular dynamics data.[100]

The bridge functions were obtained from $g_{ij}(r)$ by the definition of the exact closure (10.2.11). Their curves are displayed in Figure 10.5.2. The plot is for $-B_{ij}$ (negative bridge function) vs. r in Å. We see that they are not zero, contrary to what was stipulated by the HNC closure. In fact, the absolute values tend to increase for the like pairs as r falls below 2Å. HNC is incorrect for the molten salts studied here. Thus a need for a new closure that can yield the "true" bridge functions in order to describe correctly the structural behavior of charged species at all concentrations. We shall describe one of such efforts next.

10.6. Characterization of the Bridge Function

We observe that the bridge function (or closure relation) is a crucial component of the integral equation approach. If we have "good" bridge functions, we can get accurate structures of the ionic fluids as well as their thermodynamic properties. We have just shown that the bridge functions for molten salts are not zero. In fact for electrolyte solutions, they also differ from zero. Duh et al.[26] have determined the bridge functions for 2-2 electrolytes. Figure 10.6.1 shows $B(r)$ from MD simulations at concentrations from 0.005M to 0.020M.

We observe that the bridge functions are non-zero (while HNC insists that $B_{ij}=0$). It is also interesting to see that the like-charge B_{++} are negative, while the unlike-charge B_{+-} are positive. This behavior is contrary to the universality Ansatz of Rosenfeld[90] (which postulates that all classical fluids have similar bridge functions.) From this and other studies, it is by now clear that for electrolytes the bridge functions are non-zero and not "universal" ($B_{++} \neq B_{+-}$); and the HNC closure is only an approximation.

10.6.1. Development of a theory for the bridge functions

We develop here a general theory for the bridge function that is applicable not only to electrolytes, but also to other classical fluids. The starting point is the *singlet direct correlation function* (sdcf) $c^{(1)}$ defined for a non-uniform system (systems where there is spatial density stratification or variation). Let the nonuniformity be generated by an external one-body potential, $w(r_k)$. The system has N fluid particles. These particles also interact with a particle-particle pair potential $u(r_{ij})$. The indices ij count all the $N(N-1)/2$ pairs of i and j particles. In addition, each particle interacts with the external source potential $w(r_k)$ (which can be physically a solid wall, an electric field, or a test particle). The index k counts all N particles. The Hamiltonian of an N-body system under the influence of a wall potential and a pair potential is

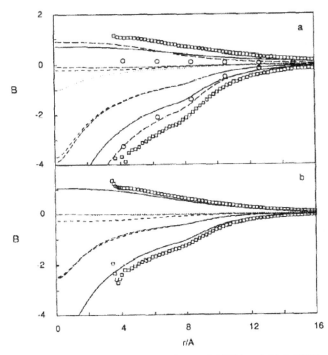

Figure 10.6.1. The bridge functions for (a) 0.005M and (b) 0.020M aqueous 2-2 electrolytes from molecular dyanmics.[26] Positive (B>0) curves are for the unlike charges. Negative (B<0) curves are for the like charges. Squares = MD data. Lines = theories (see Text). Note that for the HNC closure, $B_{ij}=0$. Clearly, the bridge functions are non-zero according to MD.

$$H_N(p^N, r^N) = \sum_{i=1}^{N} \frac{p_i^2}{2m} + V_N(r^N) + W_N(r^N) = \sum_{i=1}^{N} \frac{p_i^2}{2m} + \sum_{i<j=2}^{N} u(r_{ij}) + \sum_{k=1}^{N} w(r_k)$$

$$(10.6.1)$$

The sum of the pairwise pair interactions is V_N. The sum of all particle-to-wall interactions is denoted by W_N. The singlet direct correlation is defined herein as

$$c^{(1)}(r_1; w) \equiv \ln[\rho^{(1)}(r_1; w)\Lambda^3] + \beta w(r_1) - \beta \mu \qquad (10.6.2)$$

It is a function of the vector position r_1. The argument w represents the influence of the wall potential. This sdcf plays an important role in the theory of liquids.[59,60] First of all, all higher order ($n=2,3,..$) direct correlation functions are generated from this sdcf. The *pair* direct correlation we have use before is given by

$$c^{(2)}(r_1, r_2; w) \equiv \frac{\delta c^{(1)}(r_1; w)}{\delta \rho^{(1)}(r_2)} \qquad (10.6.3)$$

Still higher orders are defined as

$$c^{(n)}(r_1,...,r_n; w) \equiv \frac{\delta^{n-1} c^{(1)}(r_1; w)}{\delta \rho^{(1)}(r_n)...\delta \rho^{(1)}(r_2)}, \qquad n = 2,3,4,... \qquad (10.6.4)$$

This sdcf also admits a "functional" Taylor expansion: expanding around $w = 0$ (when $w = 0$, the external potential vanishes. We recover the uniform fluid):

$$c^{(1)}(r_1; w) - c^{(1)}(r_1; w = 0) =$$

$$\frac{\rho}{1!} \int d2 \; c^{(2)}(12;0)h(10) +$$

$$+ \frac{\rho^2}{2!} \int d2 d3 \; c^{(3)}(123;0)h(20)h(30) + \qquad (10.6.5)$$

$$+ \frac{\rho^3}{3!} \int d2 d3 \; c^{(4)}(1234;0)h(20)h(30)h(40) +$$

$$+ \frac{\rho^4}{4!} \int d2 d3 \; c^{(5)}(12345;0)h(20)h(30)h(40)h(50) +$$

$$+ ...$$

where the arguments ($r_1, r_2, r_3, r_4, ..., r_0$) are abbreviated to (1234...0). The argument 0 denotes the test particle position r_0. (A test particle is a pseudo-wall, a source of force for non-uniformity. It can be a spherical object or one of the fluid particles, singled out for special consideration). We *define* the bridge function B to be the sum of the expansion terms beyond the pair term $c^{(2)}$. (This is a more fundamental definition of the bridge function!)

$$c^{(1)}(r_1;w) - c^{(1)}(r_1;w=0) = \frac{\rho}{1!}\int d2 \ c^{(2)}(12;0)h(10) + B(10), \quad where$$

$$B(10) \equiv \frac{\rho^2}{2!}\int d2d3 \ c^{(3)}(123;0)h(20)h(30) +$$

$$+ \frac{\rho^3}{3!}\int d2d3 \ c^{(4)}(1234;0)h(20)h(30)h(40) + \tag{10.6.6}$$

$$+ \frac{\rho^4}{4!}\int d2d3 \ c^{(5)}(12345;0)h(20)h(30)h(40)h(50) +$$

$$+ ...$$

It can be easily shown that the left hand side is the cavity function ln$y(10)$, and the first term (on the RHS) in (10.6.6) is part of the OZ equation. Thus

$$\ln y(10) = c^{(1)}(r_1;w) - c^{(1)}(r_1;w=0) = h(10) - c^{(2)}(10) + B(10), \tag{10.6.7}$$

We have derived an "exact" closure relation in terms of the Taylor expansion involving nth order direct correlation functions. If we knew all the high order direct correlation functions $c^{(n)}$, n>2, we could obtain the "exact" bridge function. The fact is that we do not have precise means of obtaining $c^{(3)}$ and beyond. Nonetheless, we can make reasonable approximations to the higher order direct correlations. A number of attempts have been made to capture the behavior of $c^{(3)}$. One formulation is due to Barrat et al.[6] We shall briefly discuss the essentials of their approach.

Barrat et al.[6] first postulated that the triplet direct correlation can be decomposed as a product of an unknown pair function: $t(r_i,r_j)$ which is arbitrary and is to be determined by known relations

$$c^{(3)}(r_1,r_2,r_3) \equiv t(r_1,r_2)t(r_1,r_3)t(r_3,r_2) \tag{10.6.8}$$

Then from the exact sum rule on $c^{(3)}$ we know

$$\frac{\partial c^{(2)}(r_1,r_2)}{\partial \rho} = \int dr_3 \ c^{(3)}(r_1,r_2,r) = t(r_1,r_2)\int dr_3 \ t(r_1,r_3)t(r_3,r_2) \tag{10.6.9}$$

Since we can obtain the density derivative $\partial c^{(2)}/\partial \rho$ from the Ornstein-Zernike equation, we can solve (10.6.9) as a supplementary integral equation once we know $\partial c^{(2)}/\partial \rho$. This gives the solution for $t(r_i,r_j)$. The result for soft potentials (e.g., Lennard-Jones, or inverse 12th power potentials) is shown in Figure 10.6.2. We observe that the shape of the t-function resembles the curve of a direct correlation function $c^{(2)}$. We conjecture that a superposition of the pair dcf $c^{(2)}$ will probably be equally effective. (k being a constant to be determined):

$$c^{(3)}(r_1, r_2, r_3) \equiv k\, c^{(2)}(r_1, r_2) c^{(2)}(r_1, r_3) c^{(2)}(r_3, r_2) \qquad (10.6.10)$$

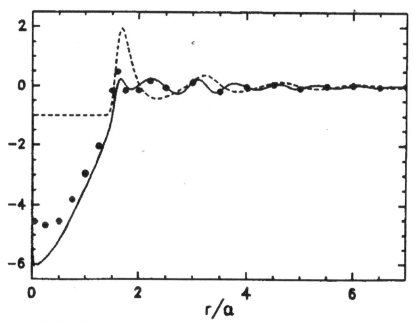

Figure 10.6.2. The test function $t(r)$ (solid line) as obtained from the Barrat-Hansen-Pastore[6] approximation(10.6.8) to the triplet direct correlation function $c^{(3)}$. The dashed line is the corresponding total correlation function $h(r)$. The condition is for a soft sphere near freezing. The filled circles are $t(r)$ for OCP (one-component plasma) close to freezing.

This observation motivates us to make the following approximation: Let the triplet direct correlation $c^{(3)}$ be represented by a product of two pair direct correlations $c^{(2)}$ and one "modification" function $F_3(r_i, r_j, r_k)$, in the spirit of (10.6.8)

$$c^{(3)}(r_1, r_2, r_3) \equiv c^{(2)}(r_1, r_2) c^{(2)}(r_1, r_3) F_3(r_1, r_2, r_3) \qquad (10.6.11)$$

$F_3(r_i, r_j, r_k)$ can be considered as defined by the above equation in terms of $c^{(3)}$ and $c^{(2)}$. Certainly, it must satisfy the symmetry of the arguments as required by $c^{(3)}$ (invariant under permutations of three arguments r_i, r_j, and r_k). With this definition, we shall develop a summation formula for the bridge function. We first examine the $c^{(3)}$ term in the bridge expansion (10.6.6). Denote it by B_3.

$$B_3(10) \equiv \frac{\rho^2}{2!} \int d2 d3 \ c^{(3)}(123;0) h(20) h(30) =$$

$$= \frac{\rho^2}{2!} \int d2 d3 \ F_3(1,2,3) [c^{(2)}(12;0) h(20)] [c^{(2)}(13;0) h(30)] =$$

$$= \frac{\rho^2}{2!} \int d2 [c^{(2)}(12;0) h(20)] \bullet \int d3 \ [c^{(2)}(13;0) h(30)] \bullet F_3(1,2,3) = \qquad (10.6.12)$$

$(apply \ mean - value \ theorem \ on \ F_3)$

$$= \frac{1}{2!} \overline{F}_3 [h(10) - c^{(2)}(10)]^2 = \frac{1}{2!} \overline{F}_3 \ \gamma^2(10)$$

To make progress, we have applied the mean-value theorem to $F_3(r_i, r_j, r_k)$ in the last equality. This gives a mean value of $\overline{F}_3$. Also we have used the OZ relation for the convolution of $h(r)$ and $c(r)$ to give the indirect correlation $\gamma(r)$. Note that due to taking of the mean values, $\overline{F}_3$ is now dependent on the temperature T and density ρ of the system $\overline{F}_3 = \overline{F}_3(T, \rho)$. Similarly, we factorize $c^{(4)}$ in terms of $F_4(r_i, r_j, r_k, r_l)$

$$c^{(4)}(r_1, r_2, r_3, r_3) \equiv c^{(2)}(r_1, r_2) c^{(2)}(r_1, r_3) c^{(2)}(r_1, r_4) F_4(r_1, r_2, r_3, r_4) \quad (10.6.13)$$

Applying the mean value theorem to $F_4(r_i, r_j, r_k, r_l)$, the second term in (10.6.6) becomes

$$B_4(10) \equiv \frac{\rho^3}{3!} \int d2 d3 d4 \ c^{(4)}(1234;0) h(20) h(30) h(40) =$$

$$= \frac{\rho^3}{3!} \int d2 d3 d4 \ F_4(1234) [c^{(2)}(12;0) h(20)] [c^{(2)}(13;0) h(30)] [c^{(2)}(14;0) h(40)]$$

$(apply \ mean - value \ theorem \ on \ F_4)$

$$= \frac{1}{3!} \overline{F}_4 [h(10) - c^{(2)}(10)]^3 = \frac{1}{3!} \overline{F}_4 \ \gamma^3(10)$$

$$(10.6.14)$$

Repeating the use of modification functions F_n to higher order $c^{(n)}s$, we can write the bridge function as an infinite series

$$B(10) \equiv \frac{\overline{F_3}}{2!}\gamma^2(10) + \frac{\overline{F_4}}{3!}\gamma^3(10) + \frac{\overline{F_5}}{4!}\gamma^4(10) + ..., \quad or$$

$$B(r) \equiv \frac{\overline{F_3}}{2!}\gamma^2(r) + \frac{\overline{F_4}}{3!}\gamma^3(r) + \frac{\overline{F_5}}{4!}\gamma^4(r) + ...,$$

(10.6.15)

The mean values $\overline{F_3}$, $\overline{F_4}$, ... are now functions of T and ρ; and $r = |\mathbf{r}_0 - \mathbf{r}_1|$ is the inter-particle distance. Let us examine the PY approximation (10.2.12) by expanding it into Taylor series

$$B(r) \cong \ln[1 + \gamma(r)] - \gamma(r) = -\frac{1}{2}\gamma^2 + \frac{1}{3}\gamma^3 - \frac{1}{4}\gamma^4 - +...(PY) \quad (10.6.16)$$

Comparison with (10.6.15) shows that the "choices" made of the mean values $\overline{F_i}$ for PY are

$$\overline{F_3} = -1, \quad \overline{F_4} = 2, \quad \overline{F_5} = -6, \quad ...etc. \quad (PY) \quad (10.6.17)$$

Thus in the PY approximation, the mean values $\overline{F_i}$ are constants and independent of the state conditions. This clearly makes it easier to use in calculations. There are other closure relations in literature. One of the more successful ones is the Verlet bridge function (α being a constant = 0.8).

$$B(r) \cong -\frac{\gamma^2}{2(1+\alpha\gamma)} = -\frac{\gamma^2}{2}\left[1 - \alpha\gamma + \alpha^2\gamma^2 - \alpha^3\gamma^3 + \alpha^4\gamma^4 - +...\right] \quad (Verlet) \quad (10.6.18)$$

Comparison with (10.6.15) gives

$$\overline{F_3} = -1, \quad \overline{F_4} = 3\alpha, \quad \overline{F_5} = -12\alpha^2, \quad ...etc. \quad (Verlet) \quad (10.6.19)$$

It is clear that the expansion of Verlet differs from the PY closure. However, by choosing $\alpha = 2/3 = 0.667$, we can match the coefficients of the first two terms. After that, the two closures start to deviate from each other. Verlet's closure has been shown for the hard sphere system to be far superior to the PY closure. Another closure of interest is the so-called ZSEP closure (the *zero-separation closure* i.e. a closure that satisfies the zero-separation theorems on the cavity functions). (α, φ, and ζ are adjustable parameters and are functions of T and ρ)

$$B(r) \cong -\frac{\zeta}{2}\gamma^2(r)\left[1 - \phi + \frac{\phi}{1 + \alpha\gamma(r)}\right] = -\frac{\gamma^2}{2}\zeta\left[1 - \phi\alpha\gamma + \phi\alpha^2\gamma^2 - \phi\alpha^3\gamma^3 + -...\right] \quad (10.6.20)$$

This closure reduces to the Verlet closure upon setting $\varphi = 1$ and $\zeta = 1$. Comparison with (10.6.15) gives the following interpretation of the mean values

$$\overline{F}_3 = -\zeta, \quad \overline{F}_4 = 3\alpha\phi\zeta, \quad \overline{F}_5 = -12\alpha^2\phi\zeta, \quad ...etc. \quad (ZSEP) \quad (10.6.21)$$

We have remarked earlier that the mean values $\overline{F}_i$ should be functions of temperature and density of the system, because the nth direct correlation functions that were being averaged were functions of T and ρ. Indeed, in the studies where the ZSEP closure was applied (to various fluids of hard spheres, Lennard-Jones fluid, and penetrable spheres), the parameters α, φ, and ζ were shown to be dependent on the state condition in order to get good results. This is consistent with the observations on the mean values $\overline{F}_i$.

10.6.2. Renormalized bridge functions

We have derived a new formula for the bridge function (10.6.15), where we have changed it from a *functional* of the total correlations $h(r)$ and the nth-order direct correlations $c^{(n)}(123...)$ to a *function* of the indirect correlation $\gamma(r)$! (i) This was done at the expense of making the coefficients $\overline{F}_i$ temperature and density dependent. (ii) In literature there have been conjectures on what should be the "correct" independent variable. In (10.6.15), the indirect correlation function $\gamma(r)$ naturally arose as the correct argument. Martynov et al.[73] suggested for this role the thermal potential $\omega(r)$ which is defined as $\omega(r) = \ln y(r)$. Others[19,26] used a renormalized indirect correlation function: γ^*

$$\gamma(r) \equiv \gamma^*(r) + \Delta\gamma(r) \quad (10.6.22)$$

$\Delta\gamma(r)$ being some known function that was often assumed to be the attractive (*att*) or long-range part of the pair potential, $\beta u^{att}(r) \equiv \beta u(r) - \beta u^{rep}(r)$ (*rep*= repulsive). The idea is to "extract" out, if possible, all the long-range (or *extra-functional*) effects from the argument of the bridge function so as to make the bridge function genuinely short-ranged and behave like a "function" (instead of a "functional"). Actual machine data[19,26] (MD or MC simulations) showed that this correction is necessary. The reason is that no simple mathematical function can capture the behavior of the modification functions $\overline{F_i}$.

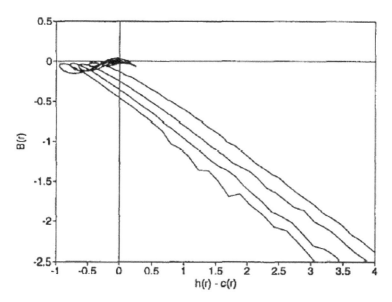

*Figure 10.6.3. The raw Monte-Carlo data for the **B-γ** plot for Lennard-Jones potential at four conditions[19] (The curves from left to right are for $\rho^*=0.85$ with $T^*=0.72$ and $\rho^*=0.8$ with $T^*=0.81$, 1.0,and 1.5.) The "function" is multi-valued— it is not a mathematically defined "function"!*

To elaborate on this non-functionality, Llano-Restrpo and Chapman[19] in 1994 used Monte Carlo simulations to determine the bridge function for the Lennard-Jones potential. They plotted the raw machine data $B(r)$ vs $\gamma(r)$. The procedure is to take a value $r=r_1$, then read both $B_1=B(r_1)$ and $\gamma_1=\gamma(r_1)$. This gives the pair (γ_1,B_1). Next, take $r=r_2$, read (γ_2,B_2). Continue with this collection (γ_3,B_3), (γ_4,B_4) ...etc. Then they plotted γ_i as the abscissa and B_i as the ordinate. This gives a ***B-γ plot***. If B is a "true function" of γ, then one would get a single-valued "function" relation between the two variables. On the contrary, they obtained a multi-valued curve (Figure 10.6.3). This indicates that the nature of a "functional" showed up, and no "function" existed to describe the relation. To extract out the r-dependence, they "renormalized" the indirect correlation γ by (10.6.22). The $\Delta\gamma(r)$ they have chosen is the attractive potential corresponding to the Weeks-Chandler-Andersen (WCA) perturbation term ϕ_{WCA}.

$$\Delta\gamma \equiv \beta\phi_{WCA}(r) = \beta u(r) - \beta u^{rep}(r) \qquad (10.6.23)$$
$$\gamma^*(r) \equiv \gamma(r) - \beta\phi_{WCA}(r) \qquad (10.6.24)$$

Then they plotted, instead of the $B(r)$ vs $\gamma(r)$, $B(r)$ vs. the renormalized $\gamma^*(r)$. The result is Figure 10.6.4.

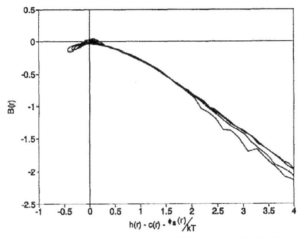

Figure 10.6.4. The B-γ^* plot in terms of the renormalized indirect correlation[19] $\gamma^*\equiv\gamma(r)-\beta\phi_{WCA}(r)$. All curves "collapse" into a single curve (within numerical scatter). The multi-values disappear. It behaves like a normal mathematical "function". (Curves are at $\rho^*=0.8$ with $T^*=0.81$, 1.0, and 1.5 and $\rho^*=0.85$ with $T^*=0.72$.)

We formalize this procedure. If we apply the split (10.6.22) to (10.6.15), we can separate out an *excess part* ΔB from the bridge function according to: $B = B^*(\gamma^*) + \Delta B(\gamma^*, \Delta\gamma)$

$$B(\gamma) \equiv B*(\gamma*) + \Delta B(\gamma*, \Delta\gamma), \qquad where$$

$$B*(\gamma*) \equiv \frac{\overline{F_3}}{2!}\gamma* + \frac{\overline{F_4}}{3!}\gamma* + \frac{\overline{F_5}}{4!}\gamma* + ..., \quad and$$

$$\Delta B(\gamma*, \Delta\gamma) = \Delta\gamma\left[\frac{\overline{F_3}}{1!}\gamma* + \frac{\overline{F_4}}{2!}\gamma*^2 + \frac{\overline{F_5}}{3!}\gamma*^3 + ...\right] + \qquad (10.6.25)$$

$$+ \Delta\gamma^2\left[\overline{F_3} + \frac{\overline{F_4}}{2}\gamma* + \frac{\overline{F_5}}{4}\gamma*^2 + ...\right] + ...$$

(Namely, we have simply substituted $\gamma \equiv \gamma* + \Delta\gamma$ into (10.6.15), expanding the powers and collecting terms). One choice for $\Delta\gamma$, among many others, could be (10.6.23)

$$\Delta\gamma \equiv \beta u^{att}(r) = \beta u(r) - \beta u^{rep}(r)$$

Other choices exist in literature.[26,27,61] According to the simulation data[19] on Lennard-Jones systems and 2-2 electrolytes,[26] B*($\gamma*$) is much better behaved than B(γ), and is short-ranged, so long we have properly selected a renormalization function $\Delta\gamma$. The term ΔB behaves like an additional "pair interaction" that modifies the pair potential $\beta u(r)$. The closure can now be written as

$$\ln y(r) = \ln g(r) + \beta u(r) = \gamma(r) + B(\gamma(r)) = \gamma* + \Delta\gamma + B*(\gamma*) + \Delta B(\gamma*, \Delta\gamma), \quad thus$$
$$\ln g + [\beta u - \Delta B(\gamma*, \Delta\gamma) - \Delta\gamma] = \gamma* + B*(\gamma*), \qquad or$$
$$\ln\left[ge^{\beta u - \Delta B - \Delta\gamma}\right] = \gamma* + B*(\gamma*),$$

$$(10.6.26)$$

We have modified the cavity function and also renormalized the bridge function. This is in the spirit of the bridge function formulations of Rogers-Young[89] and Hansen-Zerah[42] (the HMSA—Hybrid MSA closure) that have found applications in liquid theories.

To recapitulate, our derivations of the bridge function were based upon the Taylor expansion of the singlet direct correlation. We have elucidated the transformation of a bridge *functional* to a bridge *function*, by introducing the modification functions F_n. We have further given a basis for the renormalization of the argument function from γ to $\gamma*$ and expressed explicitly the function form of the renormalized potential ΔB. Once a proper $\Delta\gamma$ is chosen, we can implement the closure (10.6.26), by incorporating the formula (10.6.25) for ΔB.

The modification functions F_n are rigorously defined functions, albeit unknown. The mean-value theorem is applicable when the topological conditions are satisfied. The indirect correlation $\gamma(r)$ appears to be the proper argument function out of our development. The infinite series (10.6.15) contains all these unknown $\overline{F}_i$ coefficients. They must be determined by some means, preferably other than from their definitions. To determine them *indirectly*, we have proposed earlier the use of the thermodynamic consistencies (e.g. the Maxwell relations, pressure consistency, and Gibbs-Duhem relation) and the structural consistencies (the zero-separation theorems, the contact value theorems) plus other sum rules. The principle is, instead of calculating $\overline{F}_i$ from their definitions, we obtain indirectly a set of $\overline{F}_i$'s that happen to "*enforce*" these consistency conditions. This is called the *self-consistency principle* (to be specific, we coin the word, "*autochthony*"— a method that is independent, self-originating, and free of external influences). That is: the theory can, like boot-strapping, "improve" itself on the flight (during the numerical calculations), so that many of the consistency rules are obeyed. The bridge function is made compliant with a number of adjustable parameters. The values of parameters can be altered during the calculations. We "mold", so to speak, the bridge function during the numerical solution so as to satisfy the consistency rules. The hope is that with enough self-consistencies, $B(r)$ will behave nicely and approach the "exact" bridge function. The proof for now is in the results (*the proof of the pudding is in the eating!*)

The ZSEP closure has adjustability built-in via a number of flexible parameters (α, φ, and ζ). We have made such tests on a variety of systems to check its performance: for example, on hard spheres, hard sphere mixtures, non-additive-diameter hard sphere mixtures, hard diatomics, soft spheres and Lennard-Jones molecules, also penetrable spheres and confined fluids (via the replica OZ relation). Highly accurate answers were obtained in each case when compared to simulation. We attribute the success to (i) the self-consistency conditions, and (ii) a suitably flexible bridge function. If we examine the ZSEP closure (10.6.20), it is essentially a Padé approximant with adjustable parameters, α, φ, and ζ. The parameters are related to the mean values $\overline{F}_i$ through eq.(10.6.21). α, φ, and ζ can be varied and updated during numerical solution so that the consistency conditions can be satisfied. Since there are only three parameters, we can determine at most three $\overline{F_i}$'s, say, $\overline{F_3}$, $\overline{F_4}$, and $\overline{F_5}$. Hopefully the higher order terms $\overline{F_i}$ are small and thus negligible; or they are reproduced by the resummation (10.6.21) through the Padé formula. The ZSEP bridge has the leading term $[-(\frac{1}{2})\gamma^2(r)]$ arising from exact cluster theory. (This term is exact). The remainder is a simple Padé approximant. It interpolates between the terms $[-(\frac{1}{2})\gamma^2(r)]$ and $[-1/(1+\alpha\gamma(r))]$ by choosing a ϕ. As a matter of philosophy, we are not attached to the particular function form in (10.6.20). Any other flexible functional forms that are effective are acceptable. The major thrust is autochthony—self-generating, self-correcting through consistency.

10.7. Isothermal Compressibility and Moment Rules

Ionic distribution functions obey certain sum rules (values of integrals) due to their Coulombic nature. We introduce two moment rules (the *zeroth moment condition* and the *second moment condition*) and a relation on the isothermal compressibility. We state here without proof. (For details, see Attard.[3])

10.7.1. Isothermal compressibility

For a single salt with one cation species and one anion species, the isothermal compressibility $(\partial\rho/\partial P)_T$ is related to the zero k-values of the Fourier space total correlations

$$\frac{1}{\rho_{TOT}}\left(\frac{\partial\rho_{TOT}}{\partial\beta P}\right)_T = \tilde{h}_{+-}(k=0) = \frac{1}{\rho_+} + \tilde{h}_{++}(k=0) = \frac{1}{\rho_-} + \tilde{h}_{--}(k=0) \quad (10.7.1)$$

Note that each pair (cation-cation, anion-anion, or cation-anion) separately gives the isothermal compressibility. Also for the inverse isothermal compressibility (where $\rho_{TOT} = \rho_+ + \rho_-$),

$$\rho_{TOT}\left(\frac{\partial\beta P}{\partial\rho_{TOT}}\right)_T = \rho_{TOT} - \rho_+^2\tilde{c}_{++}(k=0) - \rho_-^2\tilde{c}_{--}(k=0) - \rho_+\rho_-\tilde{c}_{+-}(k=0) \quad (10.7.2)$$

10.7.2. The zeroth moment condition: The electroneutrality

The total correlation function has the property that the sum of all other charges around a center ion of species i will balance the amount of charge q_i of the center.

$$\sum_{k \neq i} \rho_k q_k \int d\vec{r} \, h_{ki}(r) = \sum_{k \neq i} \rho_k q_k \int_0^\infty dr \, 4\pi r^2 \, [g_{ki}(r) - 1] = -q_i \qquad (10.7.3)$$

This relation is equivalent to the electroneutrality condition given before

$$\sum_j \rho_j q_j = 0 \qquad (10.7.4)$$

Note that $q_j = z_j e$ is the Coulomb charge on ion j.

10.7.3. The second moment condition

From analyses of the Fourier space expansions of the total correlations, the following condition is derived (Stillinger and Lovett[98])

$$\frac{1}{6\varepsilon_m kT} \sum_i \sum_j \rho_i \rho_j q_i q_j \int_0^\infty dr \, 4\pi \, r^4 h_{ij}(r) = -1 \qquad (10.7.5)$$

The Fourier space expansions (in powers of the reciprocal vector k) for the correlation functions contain only even powers of k. Thus there are no odd moment conditions. These sum rules are useful in checking the validity of the correlation functions calculated by approximate theories.

Exercises:

10.1. Develop a Fortran (or C) program for solving the Ornstein-Zernike equation with the HNC closure for the soft Coulomb potential (10.3.12) at conditions: T= 298K, D= 78.358, for a 2-2 symmetric electrolyte solution at 0.005M. Plot the $g_{++}(r)$, $g_{+-}(r)$, and calculate the osmotic coefficient.

10.2. Use the Born-Huggins-Mayer potential (10.4.1) to model the molten salt NaCl at T=1165K, and ρ=0.0314 ions/$\mathring{A}^3$. Solve for the HNC and PY closures and compare with the molecular dynamics data of Tasseven et al.

10.3. Find the triplet direct correlation $c^{(3)}(1,2,3)$ by using eq.(10.5.8b) (instead of the t-function of (10.5.8)). Use the Lennard-Jones potential at the reduced temperature $T^* = kT/\varepsilon = 0.90$, density $\rho^* = \rho\sigma^3 = 0.75$. Find the value k. Repeat the solution for $T^* = kT/\varepsilon = 1.05$ and $\rho^* = \rho\sigma^3 = 0.65$. What is the value k now?

10.4. The Martynov-Sarkisov closure is of the form

$$B(\gamma) = (1 + 2\gamma)^{1/2} - \gamma - 1$$

which has been shown to be very accurate for hard spheres. Match the expansion coefficients with the modification functions $\overline{F}_3$, $\overline{F}_4$, and $\overline{F}_5$.

10.5. The Ballone-Pastore-Galli-Gazillo closure is of the form

$$B(\gamma) = (1 + s\gamma)^{1/s} - \gamma - 1$$

where $s = 15/8$. It has been shown that this closure does not give "imaginary" bridge values as the Martynov-Sarkisov closure. Match the expansion coefficients with the modification functions $\overline{F}_3$, $\overline{F}_4$, and $\overline{F}_5$.

Chapter 11

The Electric Double Layers

11.1. Introduction

Electric double layer (EDL) refers to the interface of a liquid electrolyte solution adhering to a charged solid surface (wall) as the latter is being immersed in the liquid. The counter-ions with charges opposite to the surface charge accumulate on the wall and establish an electric potential ψ in the liquid. The EDL is important in colloidal chemistry, biochemistry, cellular surfaces, and the electrochemistry of electrodes. Colloids usually have surfaces with electric charges. Electric double layers will form on these surfaces. The repulsive forces between the electric double layers keep the colloids from collapsing, enabling them to remain in solution. Proteins and cellular membranes also carry charges. For their stability and function, the EDL plays a major role. Figure 11.1.1 shows a schematic of the interface. The solid wall is here negatively charged with a *surface charge density* σ (Coulombs/cm^2). The wall, for instance for proteins, can also carry positive charges, depending on the *pH* value of the solution (i.e. the *isoelectric point:* the *pH* where the surface charge is zero and between sign changes). The counterions (oppositely charged ions, in this case, cations) will accumulate at the solid surface excluding the coions (anions with same polarity as the wall). Two liquid layers are formed. The inner layer is called the *Stern*[97] *layer* (or the *Helmholtz layer*[45]) and is composed mostly of cations. They screen the negative charges of the wall. The other ions, coions and extra counterions, due to thermal motion and entropic effects, move out into the liquid and form the outer layer, i.e. the *diffuse layer* (also called the *Gouy-Chapman layer*). The dashed line in the figure indicates roughly the limit of the Stern layer. The thickness of the Stern layer is of the dimension of a molecule (the hard core size d of a molecule is about few Ångstroms to a nanometer). The diffuse layer extends beyond the Stern boundary into the bulk liquid with thickness of the order of κ^{-1}, the reciprocal Debye inverse length.

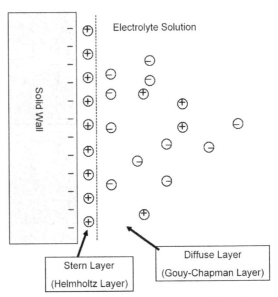

Figure 11.1.1. The electric double layer. The solid surface carries negative charges with charge density σ. The liquid side forms two layers, an inner Stern layer and an outer diffuse layer. An electric potential ψ is established in the liquid that falls off from the wall into the bulk liquid.

Several quantities are of interest in EDL: (1) the *average electrostatic potential* (AEP) ψ created in the EDL as a function of surface charge density σ, and of the distance from the wall. (2) The cation and anion density profiles in the liquid; and (3) the thermodynamic properties. For electrokinetic phenomena (such as in electrophoresis, electro-osmosis, streaming potentials, and sedimentation potentials) there is a so-called *ζ-potential*. When an electric field is applied parallel to the wall, the charged particles start to flow in the direction of the field. The ζ-potential is defined as the potential measured at the *plane of shear* (the *slipping plane* between the stagnant layer of liquid sticking to the wall and the moving part of the liquid). The plane is estimated being at a distance δ from the wall. ζ-potential has received much attention in colloid chemistry and biochemistry. We shall discuss these quantities in the following Sections. To get a physical sense of the magnitude at room conditions,[48] the AEP is about few tens of millivolts (*mV*) to few hundred *mV* in aqueous systems. The surface chare density σ is expressed in microcoulombs ($\mu C = 10^{-6}$ Coulombs) per cm^2. The thickness (extent) of the electric double layer is few nanometers (nm) into the bulk, proportional to κ^{-1} (Debye inverse length).

The methods of investigation have evolved over the last century.[96] Most theories are based on the primitive model of electrolytes (i.e. the solvent molecules are ignored as in the McMillan-Mayer picture and replaced in their place by the solvent permittivity). The earliest theories were the Gouy[39]-Chapman[20] theory and the Derjaguin-Landau[24]-Verwey-Overbeek[103] (DLVO) theory. They were based on the Poisson-Boltzmann (PB) equation that has been introduced in Chapter 4. In the mid-twentieth century, the integral equation theories[9,17] were developed for EDL based on the Ornstein-Zernike equation, or the BBGKY (Born-Bogoliubov-Green-Kirkwood-Yvon) equation[3,17]. Recently, the density functional theory has also been applied to EDL. In this Chapter, we shall introduce the PB approach. The advanced topics will be touched upon and referred for further reading to the references.[3,9,17]

11.2. The Poisson-Boltzmann Equation

The Poisson-Boltzmann (PB) equation is based on the Poisson equation of electrostatics and the Boltzmann distribution for the ions. Readers are referred to Chapter 3 for electrostatics. Here we summarize the useful equations. Let ψ be the average electrostatic potential (AEP) as defined before (Chapter 4), the Poisson equation is

$$\nabla^2 \psi(\vec{r}) = -\frac{4\pi}{\varepsilon_m} \rho_e(\vec{r}) = -\frac{4\pi}{\varepsilon_m} \sum_j q_j \rho_j^{(1)}(\vec{r}) \qquad (11.2.1)$$

where ε_m is the permittivity of the dielectric medium; ρ_e is the total electric charge density (Coulomb/volume) at the distance $\mathbf{r}$; $q_j = z_j e$ is the charge on a single ion j; $\rho_j^{(1)}(\mathbf{r})$ is the singlet probability density of ion j ar $\mathbf{r}$. The sum on j is over all ions (Na^+, Cl^-, Li^+, Br^-, etc.). For the flat wall geometry in Figure 11.1.1, the functional dependence is in the z-direction only (perpendicular to the wall). (Note: this z-coordinate is not to be confused with the valence z_j). We write (11.2.1) as

$$\frac{d^2\psi(z)}{dz^2} = -\frac{4\pi}{\varepsilon_m} \sum_j q_j \rho_j^{(1)}(z) \qquad (11.2.2)$$

The unknown quantity in (11.2.2) is the probability density of ions $\rho_j^{(1)}(r)$. According to statistical mechanics, the lowest order term in the cluster expansion of the singlet density is the Boltzmann factor

$$\rho_j^{(1)}(z) = \rho_j e^{-\beta W_j(z)} \cong \rho_j e^{-\beta q_j \psi(z)}, \qquad (approximation\ 1) \quad (11.2.3)$$

where ρ_j = number density of j (i.e. number of particles j per volume), and $W_j(z)$ is the potential of mean force. $W_j(z)$ can be set, as a first approximation, to the AEP: $W_j(z)=q_j\psi\ (z)$. With eqs.(11.2.2 & 11.2.3), we have a complete set of equations for the determination of the AEP. We need two boundary conditions for this ordinary differential equation. One is, as $z \to \infty$, the AEP should vanish, as well should its first derivative.

$$\psi(z) = 0, \quad and \quad \frac{d\psi(z)}{dz} = 0, \qquad as \quad z \to \infty \qquad (11.2.4)$$

There are a number of assumptions made in arriving at these equations. (i) The use of the Boltzmann distribution for the singlet probability density and the substitution for the potential of mean force W by the AEP. (ii) The neglect of the image charges inside the wall when going across a dielectric discontinuity. (iii) The use of the McMillan-Mayer picture, i.e. neglect of the solvent molecules. Additional approximations will be made later on in order to obtain analytical solutions. The first attempts at solution were made by Gouy[39] (1910) and Chapman[20] (1913). They solved the PB equation by assuming (iv) ions are point charges (ions of diameter zero). This, as we shall see, caused errors in the wall AEP ψ_0 (at $z=0$), because the packing of ions that do have finite volumes near the wall is limited and was not properly accounted for by point charges. Stern[98] (1924) proposed a correction to the point charge assumption by considering a hard core volume for the ions. (v) To simplify further, Gouy-Chapman linearized the exponential term in PB, similar to what Debye and Hückel did for bulk electrolytes. We have listed five simplifications above, and will mention more when they arise.

11.3. The Gouy-Chapman Theory

As alluded above, Gouy and Chapman solved the Poisson-Boltzmann equation by assuming that the ions have zero excluded volume. (In fact, they postulated a continuum picture of charges, which is equivalent to point charges).   There are two versions: (i) linearization of the exponential factor in the Boltzmann distribution (i.e. the *linear Gouy-Chapman (LGC) theory*); and (ii) keeping the exponential term, but for a symmetric electrolyte (setting $|z^+| = |z^-|$, a restriction for the sake of obtaining analytical solutions). This is the regular *Gouy-Chapman (GC) theory*.

11.3.1. The Linear Gouy-Chapman equation

The procedure of solution is similar to that for the Debye-Hückel (DH) theory, except for the geometry. After linearization of the exponent (see eq.(4.1.5)), we arrive at the expression

$$\frac{d^2\psi(z)}{dz^2} = \kappa^2\psi, \qquad where \quad \kappa^2 \equiv \frac{4\pi}{\varepsilon_m kT}\sum_j q_j^2\rho_j \qquad (11.3.1)$$

where κ is the Debye inverse length. The general solution of (11.3.1) is a linear combination of the two independent exponential functions

$$\psi = C_1 e^{-\kappa z} + C_2 e^{+\kappa z}, \qquad \frac{d\psi}{dz} = -\kappa C_1 e^{-\kappa z} + \kappa C_2 e^{+\kappa z}, \qquad (11.3.2)$$

Applying the boundary conditions (11.2.4), $C_2 = 0$. At $z=0$, the AEP should have the wall value $\psi_0 = \psi(z=0)$; $C_1 = \psi_0$.

$$\psi = \psi_0 e^{-\kappa z} \qquad (11.3.3)$$

To find ψ_0 in terms of known physics, we need some extra exact relations.

11.3.2. The sum rules for electrolytes at EDL

11.3.2.1. The electric field at the wall

We return to statistical mechanics. The average electrostatic potential $\psi(z)$ in statistical mechanics is given precisely by the expression[17]

$$\psi(z_1) = \frac{4\pi}{\varepsilon_m}\int_{z_1}^{\infty} dz_2 \,(z_1 - z_2)\left[\sum_j q_j\rho_j^{(1)}(z_2)\right] \qquad (11.3.4)$$

If we differentiate (11.3.4) with respect to z_1, we shall obtain the (negative) electric field $-E(z_1)$

$$\frac{d\psi(z_1)}{dz_1} = \frac{4\pi}{\varepsilon_m} \int_{z_1}^{\infty} dz_2 \left[\sum_j q_j \rho_j^{(1)}(z_2) \right] = -E(z_1), \quad and$$

$$\left[\frac{d\psi(z_1)}{dz_1} \right]_{z_1=0} = \frac{4\pi}{\varepsilon_m} \int_0^{\infty} dz_2 \left[\sum_j q_j \rho_j^{(1)}(z_2) \right] \tag{11.3.5}$$

We have also evaluated the gradient of ψ at the wall (i.e. at $z_1=0$). Note that the electroneutrality condition in EDL is expressed as[17]

$$\sigma = -\int_0^{\infty} dz_2 \left[\sum_j q_j \rho_j^{(1)}(z_2) \right] \tag{11.3.6}$$

Namely, the sum of all charges distributed in the double layer (per area of the wall) integrated over z_2 perpendicularly into the liquid is for all intent and purpose of neutralizing the fixed wall charge density σ (Coulomb/cm^2). Comparison of (11.3.6) with (11.3.5) gives $d\psi_0/dz$ in terms of surface charge density σ.

$$\left[\frac{d\psi(z_1)}{dz_1} \right]_{z_1=0} = -\frac{4\pi\sigma}{\varepsilon_m} \tag{11.3.7}$$

11.3.2.2. The Contact value theorem

We cite next without proof the contact value theorem relating the density distribution $\rho^{(1)}$ at the wall to the pressure P.

$$\beta P + \frac{2\pi\sigma^2}{\varepsilon_m kT} = \sum_j \rho_j^{(1)} \left(at\ z = \frac{d_j}{2} \right) \tag{11.3.8}$$

It is obtained from the BBGKY equation[9]. In the McMillan-Mayer picture, P is the osmotic pressure P^{osm}. The above relations are exact, independent of the PB equation. Next we specialize to the PB equation.

11.3.2.3. Grahame's equation

Note that the PB equations (11.2.2 & 11.2.3) say

$$\frac{d^2\psi(z)}{dz^2} = -\frac{4\pi}{\varepsilon_m}\sum_j q_j\rho_j e^{-\beta q_j\psi(z)} \qquad (11.3.9)$$

Let us define $\psi' = d\psi/dz$. We intend to integrate (11.3.9). The exponential term can be differentiated to give the relation

$$\frac{d}{dz}e^{-\beta q_j\psi(z)} = -\beta q_j\psi' e^{-\beta q_j\psi(z)}, \quad or$$

$$e^{-\beta q_j\psi(z)} = -\frac{\dfrac{d}{dz}e^{-\beta q_j\psi(z)}}{\beta q_j\psi'} \qquad (11.3.10)$$

Substituting (11.3.10) into (11.3.9), and rearranging

$$\psi'\frac{d\psi'}{dz} = \frac{4\pi kT}{\varepsilon_m}\sum_j \rho_j\frac{d}{dz}e^{-\beta q_j\psi(z)} \qquad (11.3.11)$$

Integrate with respect to dz from $z=0$ to $z=\infty$ [noting that $\psi'(z=\infty)=0$, and $\psi'(z=0)$ is given by (11.3.7)]

$$\int_0^\infty \psi'\frac{d\psi'}{dz} = \frac{1}{2}\left[(\psi')^2\right]_0^\infty = 0 - \frac{1}{2}\left[(\psi_0')^2\right] = -\frac{8\pi^2\sigma^2}{\varepsilon_m^2} =$$

$$= \frac{4\pi}{\varepsilon_m}kT\sum_j \rho_j\left[1-e^{-\beta q_j\psi_0}\right] \qquad (11.3.12)$$

Collecting the factors, we have

$$\frac{2\pi\sigma^2}{\varepsilon_m} = kT\sum_j \rho_j\left[e^{-\beta q_j\psi_0} - 1\right] \qquad (11.3.13)$$

This is Grahame's equation[40] in a general form. At low surface potential ($\psi_0\to 0$), the exponential in eq.(11.3.13) can be expanded to the second order and application of the electroneutrality rule results in

$$\psi_0 = \frac{4\pi\sigma}{\kappa\varepsilon_m} \qquad (11.3.14)$$

where κ is the Debye inverse length. This is the special form of the Grahame equation.[40]

11.3.3. Linear Gouy-Chapman equation—with boundary conditions

With the sum rule (11.3.7), we can find the wall AEP. Differentiation of ψ (11.3.3) gives

$$\frac{d\psi(z)}{dz} = -\kappa\psi_0 e^{-\kappa z},$$

$$\left[\frac{d\psi(z)}{dz}\right]_{z=0} = -\kappa\psi_0 = -\frac{4\pi\sigma}{\varepsilon_m} \qquad (11.3.15)$$

Thus ψ_0 is given by (Grahame's equation (11.3.14))

$$\psi_0 = \frac{4\pi\sigma}{\kappa\varepsilon_m}, \quad and$$

$$\psi(z)_{LGC} = \frac{4\pi\sigma}{\kappa\varepsilon_m} e^{-\kappa z} \qquad (11.3.16)$$

We have obtained the linear Gouy-Chapman (LGC) solution for the AEP. Since the singlet densities are given by the Boltzmann distribution (11.2.3), after linearization, we have

$$\rho_+^{(1)}(z) = \rho_+\left[1 - \frac{4\pi z_+ e\sigma}{\kappa\varepsilon_m kT} e^{-\kappa z}\right], \quad and$$

$$\rho_-^{(1)}(z) = \rho_-\left[1 - \frac{4\pi z_- e\sigma}{\kappa\varepsilon_m kT} e^{-\kappa z}\right], \qquad (11.3.17)$$

The linear Gouy-Chapman solution applies to the electric double layer at low surface potential, say $\psi_0 < 25 \ mV$, and low ionic strength I. Since the AEP decays exponentially, the extent of the electric double layer has a dimension of the order of a few κ^{-1}, κ is the Debye inverse length. For aqueous solutions at room temperatures, κ^{-1} is about 5Å (0.5 *nm*) at 0.1M, or about 15 Å (1.5 *nm*) at 0.01M.

11.3.4. Nonlinear Gouy-Chapman equation

In the linear Gouy-Chapman solution above, the exponential terms of the Boltzmann distributions have been linearized (by making a Taylor expansion and retaining only the linear terms). This gives poor results for strong coupling cases (high surface charges and large Debye inverse screening lengths). Attempts have been made to solve the PB equation as is— i.e. without the linearization. But there is no analytical solution for general electrolytes. Further restrictions have to be made in order to obtain analytical solutions. Thus we require that the valences be equal in magnitude (namely $z_+ =|z_-|$), i.e. *symmetrical electrolytes*. Examples are the 1-1, 2-2, or 3-3 electrolytes (such as $NaCl$, $CuSO_4$, $AlPO_4$,). Since $z_+ = |z_-| = -z_-$, $q_+ = z_+e$, and $q_-= z_- e = -q_+$, we set $q_+=|q_-|=q$. For an overall neutral electrolyte solution, $\rho_+=\rho_-=\rho$. The Boltzmann distributions can be written as

$$\sum_j q_j \rho_j^{(1)}(z) = q\rho[e^{-\beta q \psi(z)} - e^{+\beta q \psi(z)}] = -2q\rho \sinh(\beta q \psi(z)) \quad (1.3.18)$$

where *sinh* is the hyperbolic sine function. The PB nonlinear equation can now be written as

$$\frac{d^2\psi(z)}{dz^2} = \frac{8\pi q\rho}{\varepsilon_m} \sinh(\beta q \psi(z)), \quad or$$

$$\frac{d^2 \beta q \psi(z)}{dz^2} = \frac{8\pi q^2 \rho}{\varepsilon_m kT} \sinh(\beta q \psi(z)) \quad (11.3.19)$$

We have multiplied both sides by the factor βq. Let $\Phi(z)\equiv\beta q\psi(z)$. Multiply both sides by $2(d\Phi(z)/dz)$

$$2\frac{d\Phi(z)}{dz}\frac{d^2\Phi(z)}{dz^2} = 2\kappa^2 \sinh(\Phi(z))\frac{d\Phi(z)}{dz}, \quad Integrate$$

$$\left(\frac{d\Phi(z)}{dz}\right)^2 = 2\kappa^2 \cosh(\Phi(z)) + C_0 \quad (11.3.20)$$

where C_0 is an integration constant. The boundary condition is: as $z \to \infty$, $\Phi(\infty)=0$, $d\Phi(\infty)/dz= 0$, and $\cosh(0)=1$. Thus $C_0= -2\kappa^2$. Note from mathematics (the identity)

$$\sinh^2\left(\frac{\Phi}{2}\right) = \frac{1}{2}[\cosh(\Phi) - 1] \qquad (11.3.21)$$

Thus

$$\left(\frac{d\Phi(z)}{dz}\right) = \pm 2\kappa \sinh\left(\frac{\Phi(z)}{2}\right), \qquad \textit{We choose}$$

$$\left(\frac{d\Phi(z)}{dz}\right) = -2\kappa \sinh\left(\frac{\Phi(z)}{2}\right) \qquad (11.3.22)$$

We have chosen the *negative* branch because the AEP should decrease with distance z. A second integration after separation of variables gives

$$\frac{d\Phi(z)}{\sinh\left(\frac{\Phi(z)}{2}\right)} = -2\kappa\, dz, \qquad \textit{Integrate}$$

$$2\ln\left[\tanh\left(\frac{\Phi(z)}{4}\right)\right] = -2\kappa z + C_1 \qquad (11.3.23)$$

where C_1 is another integration constant. The boundary condition is: at $z=0$, $\Phi(0)=\Phi_0$. This determines C_1.

$$\ln\left[\tanh\left(\frac{\Phi(z)}{4}\right)\right] = -\kappa z + \ln\left[\tanh\left(\frac{\Phi_0}{4}\right)\right], \qquad or$$

$$\tanh\left(\frac{\Phi(z)}{4}\right) = \tanh\left(\frac{\Phi_0}{4}\right) e^{-\kappa z} \qquad (11.3.24)$$

Restoring to the AEP ψ

$$\tanh\left(\frac{q\psi(z)}{4kT}\right) = \tanh\left(\frac{q\psi_0}{4kT}\right) e^{-\kappa z} \qquad (11.3.25)$$

To find the surface charge density σ, we apply the Grahame equation (11.3.7)

$$\frac{2\pi q \sigma}{\kappa \varepsilon_m kT} = \sinh\left(\frac{q \psi_0}{2kT}\right) \tag{11.3.26}$$

The cation distribution and anion distribution are given by the equations

$$\rho_+^{(1)}(z) = \rho_+ e^{-\beta q_+ \psi(z)} = \rho\, e^{-\beta q \psi(z)}$$

$$\rho_-^{(1)}(z) = \rho_- e^{-\beta q_- \psi(z)} = \rho\, e^{+\beta q \psi(z)} \tag{11.3.27}$$

where $\psi(z)$ must be solved implicitly from (11.3.25).

We compare the linear Gouy-Chapman AEP with the nonlinear solution in Figure 11.3.1. The nonlinear Φ tracks lower than the LGC solution at the strong charge condition $\Phi_0 = 10$ (at Debye inverse length $\kappa = 1.8\text{Å}^{-1}$). This is a general trend observed for all nonlinear GC solutions.

11.4. The Stern Layer

So far we have assumed in the Gouy-Chapman theory that the ions in the electrolyte solution are point charges, without a core volume. This entails at least two anomalies. (i) There is no physical limit for the number of ions attached to the wall — there could be an infinite number of point ions at the wall if the surface charge density were high enough. In reality, sufficient but limited number of couterions will be attracted to the wall to neutralize the surface charge. They spread out in the x-y plane due to the actual size of the ions. (ii) There is no limit on how close counterions can approach the wall. Both are not non-physical. The ions have a repulsive core of the size d. Stern[98] in 1924 proposed that the first layer of ions at the wall have a size with diameter d, and they can not approach the wall closer than a distance of $z=d/2$ or radius R $(=d/2)$. Between $0<z<R$, there are no ions. This first layer of ions forms what is called the *Stern layer* (or the Helmholtz layer). Helmholtz[42] treated the double layer as a molecular condenser (a capacitor of molecular dimension), with only one layer of counterions covering the charged solid surface (thus forming a capacitor). Beyond the Stern layer, the ions remain point charges in his modified theory, and the potential again obeys the Gouy-Chapman equation at $z>d/2$. This outer layer is then called the *diffuse layer* (of Gouy-Chapman layer).

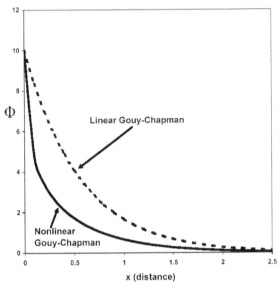

Figure 11.3.1. Comparison of the linear Gouy-Chapman equation (dashed line) with the full nonlinear Gouy-Chapman equation (solid line). The electric potential $\Phi = ze\psi/kT$ is plotted vs. distance x from the wall. Conditions: Φ_0 (at $x=0$) = 10, and Debye inverse length $\kappa = 1.8\ \text{Å}^{-1}$.

To account for the behavior in the Stern layer, a slightly modified equation is needed. We can spatially "shift" the nonlinear GC equations to accommodate ions of finite sizes. The procedure is to make the transformation—translating outward a distance of $\frac{1}{2}d$.

$$\psi^{GC}(z) \rightarrow \psi^{Stern}(z+\frac{d}{2}), \qquad \forall z > 0 \qquad (11.4.1)$$

But at $z=0$

$$\psi_0^{Stern} = \psi_{d/2}^{Stern} + \frac{4\pi\sigma}{\varepsilon_m}\frac{d}{2}, \qquad where$$

$$\beta q\,\psi_{d/2}^{Stern} = 2\sinh^{-1}\left(\frac{2\pi\beta q\sigma}{\kappa\varepsilon_m}\right) \qquad (11.4.2)$$

The subscripts on ψ are interpreted as: 0 means at $z=0$; and $d/2$ mean at $z=d/2$. Namely, we displace the wall value eq.(11.3.26) of the Gouy-Chapman AEP at z to the location $z+d/2$ for the AEP of Stern. For the Stern AEP at $z=0$, we add a correction inside the core by using the Grahame formula. Between $0<z<R$, there are no ions. There is no

contribution derived from ions in this gap to the potential except a linear drop from eq.(11.4.2).

$$\psi_0^{Stern} - \psi_{d/2}^{Stern} \cong -\frac{d\psi_0}{dz}\frac{d}{2} = \frac{4\pi\sigma}{\varepsilon_m}\frac{d}{2} \qquad (11.4.3)$$

Inside the gap $0<z<d/2$, the Stern ψ drops linearly with z. The Stern AEP for $z > d/2$ is the same as the GC AEP with z-coordinate shifted (Stern value at z) = (Gouy-Chapman value at $z-d/2$). Note the transformation (11.4.1). Outside the Stern layer, the AEP drops off exponentially. The Stern theory already improves GC in many cases. Further refinements are still possible. For real colloidal surfaces, two other important factors are missing (i) *specific ion adsorption*: certain fluid ions can be captured by the surface at specific sites. These specific adsorption sites are ubiquitous in physiological systems. The original Stern proposal[98] was formulated for this ionic adsorption. (ii) Solvent molecules which are ignored in the McMillan-Mayer treatment can exert influence in the double layer environment— this is the *solvent effects*. These effects can be accounted for to a certain extent by the *inner layer capacitance*.

Inner layer differential capacitance, C

According to electrostatics, the capacitance C is defined to be the ratio of charge Q_e (Coulomb) required per unit voltage V_e

$$Q_e \equiv CV_e, \qquad or$$
$$C = \frac{dQ_e}{dV}, \qquad or \qquad C^{-1} = \frac{dV_e}{dQ_e} \qquad (11.4.4)$$

If we apply the Stern formula, the differential capacitance C is composed of two parts: one C_i from the inner Stern layer, and C_d from the diffuse layer.

$$C^{-1} = \frac{d\psi_0^{Stern}}{d\sigma} = \frac{d\psi_{d/2}^{Stern}}{d\sigma} + \frac{d}{d\sigma}\left(\frac{4\pi\sigma d}{2\varepsilon_m}\right) = \frac{d\psi_{d/2}^{Stern}}{d\sigma} + \frac{2\pi d}{\varepsilon_m} = C_d^{-1} + C_i^{-1} \quad (11.4.5)$$

The diffuse part can be obtained by differentiating (11.4.2). The inner layer differential capacitance is of importance in many measurements on metal electrodes. (11.4.5) is a simple formulation of the differential capacitance. In real systems, the C_i dependence on σ is much more complicated.

11.5. The Zeta Potential, ζ

A charged particle in a solution under the influence of an applied electrical field E will migrate in the direction of the field. This behavior is called *electrophoresis*. It is an electrokinetic phenomenon. In an electric double layer, when the electric field is applied parallel to the wall, the mobile charges start to flow in the same direction. Since the ions in the Stern layer are held tightly, they tend to stay fixed. Into the diffuse layer, ions are moving. The division between the two regimes is called the *slipping plane* (or *plane of shear*). See Figure 11.5.1. And the average electrostatic potential ψ_ζ at this plane ($z=z_\zeta$) is called the ζ-potential.

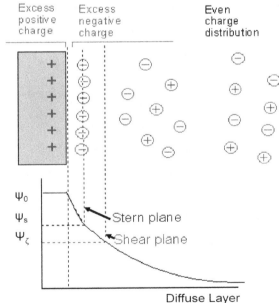

Figure 11.5.1. Schematic of the shear plane (slipping plane) relative to the electric double layer. The wall has embedded positive charges. The first layer of counterions forms the Stern layer with Stern potential ψ_s. The shear plane is somewhere outside the Stern layer. The zeta potential is indicated as ψ_ζ.

There have been from early on different conjectures as to where this slipping plane should be located. Since the PB based theories are equilibrium theories and the plane of shear is related to fluid dynamics (electrokinetics), the two branches of physics do not mesh.   It is generally accepted that the slipping plane is outside the Stern layer, somewhere in the diffuse layer.   Different electrokinetic experiments (electro-osmosis, streaming potential, and sedimentation potential) have given different values of the slipping plane and thus the ζ-potential. The situation is now improved since lately the molecular simulations have advanced considerably and one can simultaneously probe these two braches of physics (equilibrium and dynamics) directly.

The *electrophoretic mobility*, u_{EM}, is defined as the linear velocity v (cm/s) of ions per strength of electric field E (or potential drop, volt/cm).

$$u_{EM} \equiv \frac{v}{E} = \frac{velocity}{Electric\ field} \qquad (11.5.1)$$

To relate the electrophoretic mobility to experimental data, one often uses the Smoluchowski equation[71]

$$u_{EM} = \frac{\zeta \varepsilon_m}{\eta} \qquad (11.5.2)$$

where η is the viscosity (in micropoises), ε_m, the permittivity of the medium, and $\zeta = \zeta$-potential (*mV*). This equation is approximate, when only the electric driving force and frictional force are considered. It is applicable at low ζ, large colloidal particle size, and high ionic strength (e.g. $\kappa d > 100$). The electric current I_e is given[71] by

$$I_e = -\frac{\zeta \varepsilon_m}{\eta} A f_0 \qquad (11.5.3)$$

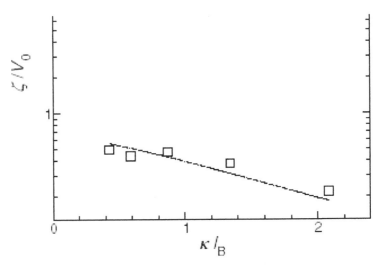

Figure 11.5.2. The square symbols show the ζ- potential measured in MD as a function the screening factor κB (B= Bjerrum length). The ζ- potential is normalized by the surface AEP ψ₀ obtained from the Poisson-Boltzmann equation at given κ and σ. The dashed line is the AEP ψ calculated by PB at z_s (at the slipping plane, as determined by MD). The position z_s of the plane of shear does not vary significantly with salt concentration.

where A is the cross-sectional area, f_0 the applied force. The variables I_e, and η can be simulated in a molecular dynamics (MD) simulation, while ε_m and f_0 can be prescribed. Precisely such a simulation has recently carried out by Joly *et al.*[50] They used a Lennard-Jones (LJ) solvent and LJ+Coulomb micro-ions in a Poiseuille flow between two walls composed of fcc lattices of solid LJ atoms. The Bjerrum length *B* was chosen to be 0.7 nm. The ionic strengths simulated were between 0.01M and 1M. The surface charge density is $-0.2e/\sigma_{LJ}{}^2$, where σ_{LJ} is the LJ size parameter (~3.4Å). The ζ-potential was calculated from (11.5.3) by MD. Their results (see Figure 11.5.2) show that the MD values of ζ (from (11.5.3)) match well with the Stern theory $\psi_{d/2}$ from (11.4.2). In other words, the slipping plane (from MD) coincides reasonably well with the Stern layer thickness for this particular EDL. (We remark that their goal of the molecular simulation was to determine the effect of boundary slip on the ζ –potential. In fact, a strong effect was detected when the fluid molecules move slipping past the solid wall.)

11.6. Beyond the Poisson-Boltzmann Theory

The Poisson-Boltzmann based theories are not the only theories for the electric double layers. Due to the introductory nature of this book, we have not delved into the rich literature since the 1980s on the integral equations and density-functional theoretical developments, which at present are the more viable and theoretically sound approaches. Recall that we have listed as least five approximations in the beginning of this Chapter regarding the Poisson-Boltzmann theory. These simplifications can be removed one by one in the statistical mechanical approaches. These new methods reflect more physical realism, freed from previous naïve assumptions. Mindful of the space and time limts, we have only looked at EDL on "flat surfaces". Other geometries (e.g. spherical and cylindrical EDL's) are probably more important, and have been investigated abundantly in literature. We shall touch upon the EDL of spherical geometry when neede. Next, we describe without derivation other modern approaches for the EDL: for example, the Ornstein-Zernike (OZ) based integral equations;[59] the BBGKY hierarchy;[8] the WLMB (Wertheim[110]-Lovett-Mou-Buff[70]) approach; and the Kirkwood hierarchy.[17,79,80]

11.6.1. Ornstein-Zernike based integral equations

In contrast to the homogenous-fluid Ornstein-Zernike integral equations discussed in Chapter 10, the nonuniform form of OZ (nonuniform because of the existence of the charged wall that imposes a potential and causes a density stratification along the z-direction) assumes the form[46,82]

$$h_j^w(1) - c_j^w(1) = \sum_k \rho_k \int d2 \, c_{jk}^B(r_{12}) h_k^w(2) \qquad (11.6.1)$$

where j,k = ion of species j or k; the superscripts w denotes wall (i.e. the non-uniformity) quantities, and B denotes bulk (uniform) quantities. h and c have their usual meanings the total and direct correlations, respectively. The arguments, $1 = \mathbf{r}_1$, and $2 = \mathbf{r}_2$, $d2 = d\mathbf{r}_2$. To solve this equation, we need two closures: one for the bulk quantities, and another for the wall (nonuniform) quantities. The HNC or MSA closures have been used (HNC was used for both singlet (w) and bulk (B) correlations; and MSA for the bulk correlations(B)). The ions now have their hard core sizes explicitly taken into account. McMillan-Mayer picture is again assumed. The results[17] are shown in Figure 11.6.1.

11.6.2. The BBGKY hierarchy

The BBGKY hierarchy[17] is based on the force balances between successive higher order density functions. The singlet density $\rho^{(1)}$ is related to the pair density $\rho^{(2)}$ via a force balance, and the pair density to the triplet density $\rho^{(3)}$ via another force balance, etc., generating a sequence of equations progressing to densities of higher and higher orders. The hierarchy is exact but not easily soluble. Approximate "closures" will have to be formulated to "truncate" the lower members of the hierarchy from the higher ones in order to obtain numerical solutions. The first member is obtained by taking the gradient of the singlet density and applying its definition to get

$$-\frac{\partial}{\partial r_1}\rho_j^{(1)}(1) = \rho_j^{(1)}(1)\frac{\partial}{\partial r_1}\beta u_j^{(1)}(1) + \sum_k \int d2 \,\rho_{jk}^{(2)}(1,2)_{12}\frac{\partial}{\partial r_1}\beta u_{jk}^{(2)}(1,2) \quad (11.6.2)$$

where $u^{(1)}$ and $u^{(2)}$ are the singlet and pair potentials. When this equation is coupled with the Poisson equation, we have a closure and the two equations can be solved. The results are shown as item "B" in Figure 11.6.1

11.6.3. The Wertheim-Lovett-Mou-Buff equation

There are several types of equations similar to the BBGKY hierarchy. One of these is the WLMB equation.[70,110] The other is the Triezenberg-Zwanzig equation.[115] All are exact until approximations are made. If the gradient of the singlet direct correlation $c^{(1)}$ is taken, we can derive

$$-\frac{\partial}{\partial r_1}\rho_j^{(1)}(1) = \rho_j^{(1)}(1)\frac{\partial}{\partial r_1}\beta u_j^{(1)}(1) - \rho_j^{(1)}(1)\sum_k \int d2 \,c_{jk}^{(2)}(1,2)\frac{\partial}{\partial r_2}\rho_k^{(1)}(2) \quad (11.6.3)$$

This equation has been studied by Plischke and Henderson.[85] The HNC2 closure was used in their work.

11.6.4. The Kirkwood hierarchy

The first member of the Kirkwood hierarchy[17,79,80] takes the form

$$\ln g_j^{(1)}(z_1) - \ln g_j^{(1)}(z_1 \mid \lambda = 0) =$$

$$= -\beta q_j \psi(z_1) - \beta q_j \int_0^1 d\lambda \left[\phi_j(z_1 \mid \lambda) - \phi_j(z_1 = \infty \mid \lambda) \right] \qquad (11.6.4)$$

where λ is the charging parameter $0<\lambda<1$; $g_j^{(1)}(z)$ is the singlet correlation $= \rho_j^{(1)}(z)/\rho_j$; and ϕ_j is a potential to be determined by a closure, such as Loeb's closure.[67] Theories based on this equation are called the modified Poisson-Boltzmann (MPB) equations.[79,80] There are many versions of MPB: from the MPB1, MPB2, ... , to MPB5. (see Bell and Levine[7]).

Figure 11.6.1 presents a comparison of the various theories[17] with Monte Carlo simulation data for a 1-1 electrolyte. The index "G" is the modified Gouy-Chapman (MGC) theory; "B" the BBGKY theory; and "H" the HNC/HNC based OZ integral theory; (HNC/HNC means that the wall closure is HNC, the bulk closure is also HNC). "E" is a modified BBGKY theory. It is seen that the MGC performs reasonably well at low molality (~0.01M) and low surface charge (σ reduced <0.1). The HNC/HNC theory is reasonable at a bit higher surface charges, but deviates abnormally at high σ. The BBGKY is good at σ up to ~2.2. Its modification "E" performs much better. Overall, it is agreed that the MGC theory is viable at low molalities and high temperature (or equivalently, solvents with high permittivity). Even for the more severe cases, the MGC is not unreasonable. Although MGC is based on a semi-continuum model of electrolytes, it must have had very fortunate cancellation of errors. This explains its wide acceptance in chemical and biochemical communities.

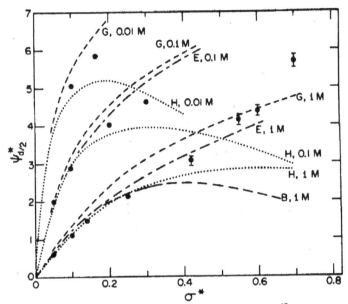

Figure 11.6.1 Comparison of $\psi_{d/2}$ from various theories[17] with the Monte Carlo (MC) simulation (black dots) at different molalities 0.01M, 0.1M. and 1M for a 1-1 electrolyte.[17] The index "G" is the modified Gouy-Chapman theory; "B" the BBGKY theory; and "H" the HNC/HNC from the OZ theory; "E" is a modified BBGKY theory.

11.7. The DLVO Theory

A popular theory of the interaction between colloids and their stability is the *Derjaguin-Landau[24]-Verwey-Overbeek[103] (DLVO) theory* developed in the 1940s. Three types of interaction between a pair of colloids are assumed: (i) a hard core repulsion, (ii) an electrostatic interaction, usually repulsive, between the electric double layers formed around a pair of colloid particles; and (iii) a short-range dispersion attraction between the colloids. It is customary to assume for simplicity that the shape of colloids is spherical. Figure 11.7.1 is a schematic of the interactions between two spherical colloids. DLVO adopts (i) the average electrostatic potential ψ (AEP) for the EDL from the Poisson-Boltzmann (PB) equation; (ii) a Hamaker dispersion potential for the attraction, and (iii) a hard sphere interaction for the colloidal cores.

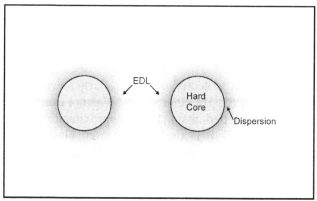

Figure 11.7.1. Two colloidal spheres with their electric double layers (EDL). The interaction is composed of three contributions (from inside out): (i) a hard core repulsion; (ii) an attractive Hamaker dispersion interaction; and (iii) the average electrostatic potential ψ from the Gouy-Chapman theory.

Before continuing with the conventional approach, let us examine some exact concepts developed from the liquid state theory.

11.7.1. The potential of mean force (PMF)

We have introduced the potential of mean force (PMF) in Section 11.2 in connection with the density profiles near a wall. (See eq.(11.2.3)) Here we define it for a pair of spherical colloid particles in solution in terms of the pair correlation function $g^{(2)}(r)$

$$W(r) \equiv -kT \ln g_j^{(2)}(r) \qquad (11.7.1)$$

This PMF is commonly accepted in colloids literature as the "actual" interaction between colloidal particles. Below we attempt to interpret it in terms of rigorous statistical mechanics. It is **not** exactly the pair potential. There are other correlation effects at play. To show this, we cite the closure relation and the zero-separation theorem.

11.7.1.1. The closure relation

As shown in Chapter 10, the closure introduces the bridge function B.

$$\ln g_j^{(2)}(r) = -\beta u(r) + h(r) - c(r) + B(r) \qquad (11.7.2)$$

In terms of the bridge function, the PMF is

$$\beta W(r) = -\ln g_j^{(2)}(r) = \beta u(r) - h(r) + c(r) - B(r) \qquad (11.7.3)$$

Namely, it is the pair potential $u(r)$ plus terms that arise from the correlations (h, c, and B). (In other words, W is not exactly the bare pair potential u). In hypernetted-chain (HNC) approximation ($B=0$), thus $\beta W \approx \beta u + h - c$. (This PMF is the pair potential plus some correlation effects). For conditions where $h=c$ (as shown in cluster series, this happens when the fluid density ρ is very low), then the PMF becomes the pair potential.

11.7.1.2. The zero-separation theorem

Zero-separation theorems[60] relate the values of the correlation functions when two molecules "coincide" (i.e., when the separation distance r between a pair of molecules = 0, as the pair interaction between them is set to zero) to the thermodynamic properties of the liquid in question. The theorems in different forms apply to a number of correlation functions, including the cavity function $y(r=0)$ and the indirect correlation $\gamma(r=0)$. For the cavity function, they read

$$\ln y(r=0) = -\beta \mu_2 + 2\beta \mu_1 \qquad (11.7.4)$$

In words, the logarithm of the cavity function is the difference between the chemical potentials $2\beta\mu_1$ for inserting two monomer spheres into the fluid (thus $2\beta\mu_1$,) and the chemical potential $\beta\mu_2$ of inserting a dimer (where μ_1 is the monomer chemical potential, and μ_2 the dimer chemical potential). The dimer is composed of two monomers fused together. Let us call this difference the *extra free energy* term. Using the definition of the cavity function $y(r) \equiv g^{(2)}(r)exp[\beta u(r)]$, we can show

$$W(L) = u(r) + \mu_2(L) - 2\mu_1 \qquad (11.7.5)$$

The argument L is the "bond length" of the dimer molecule. Thus the PMF is the sum of the pair potential and the "extra free energy". In the event, for fluids or fluid conditions, that the extra free energy is zero, μ_2

$\approx 2 \; \mu_1$ (for example, in very dilute fluids), the PMF has the value of the pair potential.

11.7.2. The DLVO interaction potentials

As described above, the DLVO interaction $W_{Colloid}$ between two colloidal spheres is composed of the hard sphere repulsion W_{HS}, the double layer AEP W_{PB} from the GC theory, and the Hamaker dispersion energy $W_{Hamaker}$. We use the potentials of mean force to represent their interactions.

$$W_{Colloid}(r) = W_{HS}(r) + W_{PB}(r) + W_{Hamaker}(r) \qquad (11.7.6)$$

The hard spheres have a diameter d_p or radius R_p (i.e. $d_p = 2R_p$).

$$W_{HS}(r) = \infty, \qquad r < d_p \qquad (11.7.7)$$

We note that for proteins, there is a layer, of thickness δ, of solvent molecules that attach themselves tightly to the surfaces, thus enlarge their effective diameter. δ is of the order of 3Å, i.e. approximately the size of a water molecule. d_p should include this increment δ.

The AEP based on the GC theory for *two charged spheres* is well-known. We present it without proof

$$W_{PB}(r) = \frac{q^2}{\varepsilon_m (1 + \kappa R_p)^2} \frac{e^{-\kappa(r - d_p)}}{r}, \qquad r > d_p \qquad (11.7.8)$$

where κ is the Debye inverse length. The Hamaker interaction is due to the dispersion forces between the two colloids (i.e. induced dipole interactions)

$$W_{Hamaker}(r) = -\frac{A_H}{6} \left[\frac{2}{(r/R_p)^2} + \frac{2}{(r/R_p)^2 - 4} + \ln \frac{(r/R_p)^2 - 4}{(r/R_p)^2} \right], \qquad r > d_p \quad (11.7.9)$$

where A_H is Hamaker constant. A_H determines the strength of the colloid-colloid dispersion interaction. Figure 11.7.2 shows the relative magnitudes of the interactions. At 5 nm, we detect the presence of a secondary minimum. As the salt concentration is increased, the first peak gradually diminished (due to the screening of the surface charges), and the combined potential becomes more and more attractive (see the progression of curves in the inset from *a, b, c, d,* to *e*). As curve *e* is approached, the interaction between the colloids is purely attractive, and the colloids, due to attraction, will become unstable and coagulate! The DLVO potential has been fitted to actual proteins (colloids). For examples,[22] lysozyme at pH= 4.2, in 0.1M sodium acetate solution, A_H = 7.7 kT, at *q*= 6.4e. For bovine serum albumin, A_H = 3 *kT*; and for subtilisin A_H =5.1 *kT*.

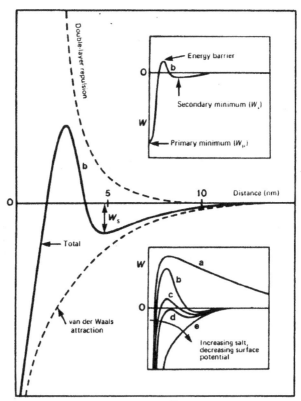

Figure 11.7.2. The DLVO interaction.[48] The y-axis is the interaction energy W(r). The x-axis is the separation distance r (in nm) measured from the surface of the hard core. The EDL interaction is usually repulsive. The dispersion force (the van der Waals interaction) is attractive. The combined interaction produces a peak (line b) that gives an additional repulsion at longer distances.

11.8. Beyond the DLVO Theory

Recent developments[13,69,92] in protein crystallization and protein phase diagram studies have prompted the search for a more accurate molecular basis to interpret the data. Figure 11.8.1 shows the crystals formed from catalase and myoglobin at *pH* 7 in ammonium sulfate solution.[28] Small angle neutron scattering (SANS) experiments[69] have been carried out for small protein molecules, such as Cytochrome C (15 x 17 x 17 Å) and lysozyme (22.5 x 15 x 15Å).

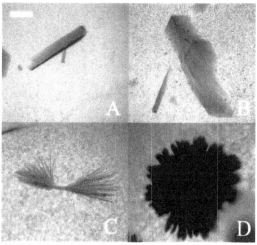

Figure 11.8.1. (A,B) Crystals of catalase formed in ammonium sulfate at pH 7, 5 mM bis-tris. (C,D) Crystals of myoglobin formed in ammonium sulfate at pH 7, 5 mM bis-tris. The scale bar represents 50 μm. (Dumetz et al.[28] 2007).

To interpret the structure factors $S(q)$ obtained from SANS, it was found that a potential as a combination[13] of two Yukawa terms best represented the data. This two-Yukawa potential consists of a short-range attraction and a long-range repulsion. Observation of a peak in $S(q)$ at small q indicates the formation of clusters of proteins at larger separations in the r-space (see Figure 11.8.2). Stradner et al.[99] in 2004 used SANS and SAXS (small-angle X-ray scattering) to study lysozyme protein solutions at 25°C and concentrations at 254 mg/cc (filled circles) and 169 mg/cc (open circles) (Figure 11.8.3). Both scattering experiments show the appearance of a small-q peak in $S(q)$ at $q_c \sim 0.78$ nm^{-1}. The interpretation is that the monomer proteins aggregate into many clusters in the solution encouraged by the strong attractive forces at short range but prevented from continued growth due to the long-range repulsion. These clusters interact to yield the small-q_c peak (the cluster-cluster interaction is shown in the illustration as the large double-arrow between a pair of cluster-sized aggregates). The large-q_m (monomer) peak of $S(q)$ (at $q_m \sim 1.3$ nm^{-1}) is due to the normal interactions between the nearest-neighbor monomer proteins, as indicated by the small double-arrow.

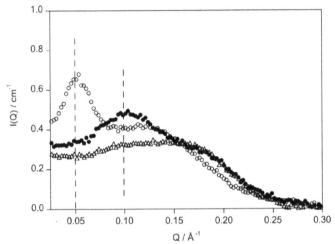

Figure 11.8.2. Experimental[69] small angle neutron scattering (SANS) intensity distribution I(Q) from Cytochrome C protein solutions of volume fraction 0.4 with and without salts. pD=11 Symbols: $\triangle$= no salt. $\bullet$= 1.9M NaCl. $\circ$= 1.2M NaSCN. In addition to the peak at large Q, another peak appears at smaller Q (at 0.10 Å^{-1} for NaCl, and 0.05 Å^{-1} for NaSCN). The large-Q peak is due to the interaction of normal colloidal proteins. The small Q peaks indicate formation of long-range cluster-like structures. Additional interaction is at play. It is suggested that a third Yukawa term can produce this effect. (Lonetti[69] 2004).

Thus the cluster formation can be engendered by the combined short-range attraction and the long-range repulsive forces.[92] The two-Yukawa potential has a hard-sphere core of diameter σ, plus two Yukawa terms: one is designed to produce the short-range attraction, and the other the long-range repulsion

$$\beta u_{2\,yukawa}(r) = \infty, \qquad r < 1$$

$$\beta u_{2\,yukawa}(r) = -K_1 \frac{e^{-z_1(r-1)}}{r} + K_2 \frac{e^{-z_2(r-1)}}{r}, \qquad r > 1 \tag{11.8.1}$$

where K_1, K_2, z_1, and z_2 are the Yukawa parameters. The K's are in units of kT. The inter-particle separation r is normalized by the length parameter σ ($r = r'/\sigma$). This two-Yukawa potential has a short-range attraction (with strength K_1), and a long-range repulsion (with strength K_2).

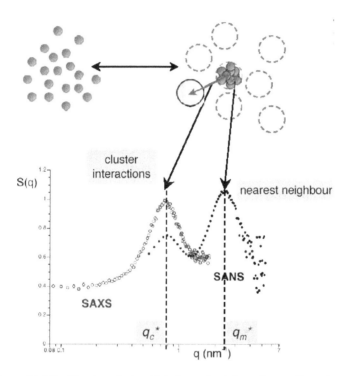

Figure 11.8.3. Structure factor $S(q)$ determined from the small angle neutron scattering (SANS) and small angle X-ray scattering (SAXS) experiments for lysozyme protein solutions at 254 mg/cc (filled symbols) and 169 mg/cc (open symbols) at 25°C. (Stradner[99] 2004).

Figure 11.8.4 shows the curves of the two-Yukawa potential at different values of K_1. We see a sharp attractive well near the hard core, then a repulsive peak at larger r-distances.

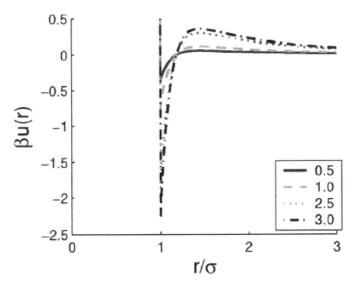

Figure 11.8.4. The two-Yukawa potential with variable K_1—the attractive strength. At high $K_1 \sim 3.0$, we see the formation of a long-range repulsion. (Broccio[13] 2006).

To describe the Cytochrome C protein structure factor $S(q)$, the following Yukawa parameters[69] were proposed (See Table 11.8.1).

Table 11.8.1. The Two-Yukawa Potential Parameters[61] for Cytochrome C*

	σ (Å)	K_1	z_1	K_2	z_2
NaCl	14.7	7±3	7±3	0.11±0.02	0.2±0.01
NaSCN	14.6	8±5	10±6	0.33±0.02	0.35±0.01
Na2SO4	14.7	4±5	10±10	0.24±0.02	0.27±0.02

*pD=11, 1% volume fraction. Salt is at 1M.

The integral equation mean spherical approach (MSA) has been solved for the Yukawa fluids and analytical formulas were given for its structure (Hoye and Blum[44] 1977). The structure factors $S(q)$ can be calculated from the pair correlations $g^{(2)}(r)$ via the well-established Fourier transform.

$$\rho\left[g^{(2)}(r)-1\right] = \frac{1}{2\pi^2 r}\int_0^\infty dq\, q[S(q)-1]\sin(qr) \qquad (11.8.2)$$

or equivalently

$$S(q) = 1 + \rho\tilde{h}(q) = \frac{1}{1 - \rho\tilde{c}(q)} \qquad (11.8.3)$$

where $\tilde{h}(q)$ and $\tilde{c}(q)$ are the Fourier transforms of $h(r)$ and $c(r)$, the total and direct correlations functions, respectively. This MSA theory was used to find the parameters of two-Yukawa potential, by fitting to the SANS $S(q)$ data. The resulting parameters are registered in Table 11.8.1.

From these developments, it is clear that there is an alternative (i.e. the integral equations) to the PB based theories for interpreting colloidal data. In the two-Yukawa approach, aside from the hard core repulsion, the interaction of the electric double layers, if any, is subsumed in the repulsive second Yukawa term (which is of precisely the same form as the Debye screened potential). The dispersion force—a short-range attraction is accounted for by the first Yukawa term (eq.(11.8.1)). Thereby the DLVO potential shape in Figure 11.7.2 is closely reproduced by the two-Yukawa potential (Figure 11.8.4). (Note to produce the second minimum in DLVO, a third Yukawa term can be added to eq.(11.8.1) without much effort).

Additional factors affecting the protein-protein interaction are now well documented[22]: (i) the anisotropy of the protein molecules (non-sphericity and non-central forces), (ii) the specific ion adsorption on protein surfaces, and (iii) solvent or solvation effects due to solvent-protein interactions. These effects have not been discussed here, and are referred to recent reviews[22,69].

Exercises:

11.1. For hard ions (cations and anios) of unequal size, Valleau and Torrie have solved the Poisson-Boltzman equation (J. Chem, Phys. 76, 4623 (1982)). Discuss this new solution: how do their results differ from the *restrict primitive model* discuss in this chapter?

11.2. Asakura and Oosawa discussed the depletion forces between hard bodies immersed in macromolecules (J. Chem. Phys. 22, 1255 (1954)). How do they compare with the DLVO potential?   In what regime do they match each other (DLVO and the Oosawa potentials)?

11.3. For the DLVO potential exhibited in Figure 11.7.2 can you find a three-Yukawa potential (potential with three Yukawa terms) to match qualitatively the short-range attraction, mid-range repulsion, and long-range attraction?

Chapter 12

Application:
Absorption Refrigeration with Electrolytes

12.1. Introduction

Refrigeration is achieved by an engine running through a cycle with a working fluid undergoing Joule-Thomson expansion. In vapor compression refrigeration (the conventional cycle), chlorofluorocarbons (CFC's) are used as working fluids. In absorption refrigeration, an absorbent together with an absorbate is used as a pair of working fluids. For example, the water-ammonia pair can be used in absorption refrigeration: water acts as the absorbent (to absorb ammonia), and ammonia is the refrigerant (absorbate) that undergoes the Joule-Thomson expansion. It has been recognized since 1973 that the CFC's are ozone-depleting chemicals and are not environmentally friendly. In 1987, the Montreal Protocol was signed by the world community to phase out the use of CFC's in refrigeration (by 1996 for CFCs 11, 12, 113, 114, and 115, and by 2030 for all HCFC's (hydrochlorofluorocarbons) which are considered less active.) There is urgent need to find alternative working fluids. One class of possible candidate is the electrolyte solutions. The ammonia-water pair first used in 1920s (the Corsley Iceball) is an example of electrolyte working fluid. Pairs of modern working fluids under current study are water-lithium bromide-ethylene glycol and ammonia-sodium thiocyanate systems.

12.2. The Absorption Refrigeration Cycle

The absorption cycle (AbC) can be contrasted with the conventional vapor compression cycle (VCC) in Figure 12.2.1 (a) and (b). The VCC uses four unit operations: 1. expansion, 2. evaporation, 3. compression, and 4. condensation. The AbC also uses four unit operations: 1. expansion (evaporation), 2. absorption, and 3. heating (or generation), and 4. condensation. The essential difference is that in VCC, a pump or a compressor driven by an electric motor is used to pump the low pressure spent gas (say, the CFC's) from the exit of the evaporator to a high pressure gas (from Point 4 to Point 1); while in AbC, the spent gas (say, ammonia) exiting the evaporator is "absorbed" by the absorbent (water) in an absorber (from Point 4 to Point 5). The liquid solution is then pumped to the generator (heater) for further regeneration to a high pressure gas (from Point 9 to Point 1).

(a)

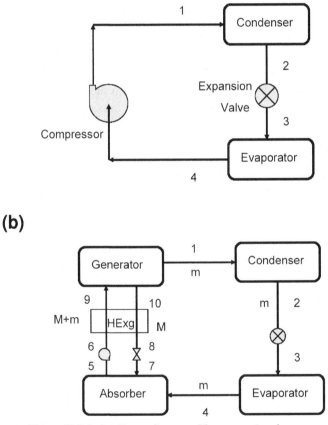

(b)

Figure 12.2.1. (a) Upper diagram--The conventional vapor compression cycle with expansion valve, evaporator, compressor, and condenser. (b) Lower diagram-- the absorption cycle with evaporator, absorber, generator, and condenser.

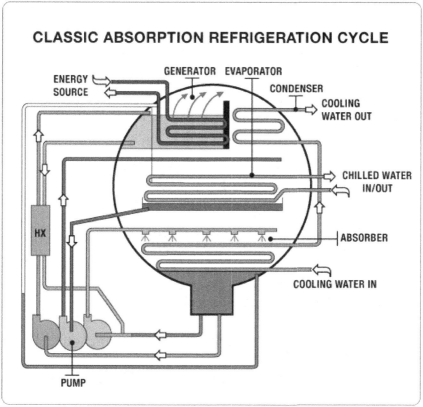

Figure 12.2.2. An integrated machine design for absorption refrigeration. A cylindrical body contains all the unit operations in a contiguous fashion.

We give a more detailed description of the absorption cycle with the water (refrigerant)-lithium bromide (absorbent) working fluid. Refer to Figure 12.2.1(b) (1) A dilute solution of LiBr-in-water in the generator is heated (Q_G) so that a high pressure water vapor is issued at Point 1 (mass flow rate = m kg/hr). The LiBr solution left in the generator becomes concentrated in the process. (2) This vapor enters the condenser and condenses into a liquid (Point 2). (3) The liquid expands and cools in the expansion valve (Joule-Thompson effect) to a two phase mixture (Point 3). (4) The mixture evaporates in the evaporator and absorbs heat Q_E. This step creates the intended cooling. (5) The low pressure vapor enters the absorber (Point 4) where a stream M kg/hr of concentrated LiBr solution (Point 7) coming from the generator enters the absorber and absorbs the water vapor from 4. (6) The resulting dilute LiBr solution exits the absorber at Point 5, and is pumped back to the generator (Point 9). The flow rate is the combined flow $m+M$ at 9. The dilute solution stream (Points 6 & 9) exchanges heat with the concentrated solution (10, 8, & 7). This takes place in the heat exchanger (*HExg*). The cycle restarts from Step (1).

The above cycle is called a *single-effect cycle*, because it has only one generator. If a second generator is added, the cycle will be called a *double-effect cycle*.

12.3. The Energy Balances in the Absorption Cycle

The energy balance is based on the conservation principle:

$$[Accumulation] = [In] - [Out] + [Source]$$

Given a system (the condenser, the evaporator, the absorber, or the generator), over a time period dt, the change of the internal energy dU in the system is related to the inlet and outlet specific enthalpies h_i (Btu/lbm or kJ/kg) as

$$d[U]_{system} = [h]_{in}\,\dot{m}_{in}dt - [h]_{out}\,\dot{m}_{out}dt + \dot{Q}_{in}dt - \dot{W}_{shaft}dt \qquad (12.3.1)$$

where Q and W are the heat input and (shaft) work output.

12.3.1. The individual equipment energy balance

We apply the conservation principle to the operating units of the AbC. At steady state, $d[U] = 0$.

The condenser

$$0 = (h_1 - h_2)\dot{m} + \dot{Q}_c \qquad (12.3.2)$$

The expansion valve

$$0 = (h_2 - h_3)\dot{m} \qquad \text{(Joule-Thomson Isenthalpic Expansion) (12.3.3)}$$

The evaporator

Note that only the liquid stream in the evaporator does the cooling. The vapor stream that enters Point 3 is shunted. However, the vapor rejoins the exit stream at Point 4. Let the quality (the fraction of vapor) of stream 3 be y.

$$0 = (1 - y)[h_3^L - h_4]\dot{m} + \dot{Q}_E \qquad (12.3.4)$$

The absorber

$$0 = \dot{m}h_4 + \dot{M}h_7 - (\dot{m} + \dot{M})h_5 + \dot{Q}_A \qquad (12.3.5)$$

The generator

$$0 = (\dot{m} + \dot{M})h_9 - \dot{M}h_{10} - \dot{m}h_1 + \dot{Q}_G \qquad (12.3.6)$$

12.3.2. The coefficient of performance, COP

The efficiency of a refrigeration cycle is measured by the *coefficient of performance* (COP) to be defined below. It is the ratio of the heat removed from the evaporator (cooling)— i.e. the benefit over the heat supplied in the generator (heating) — the cost (i.e. the ratio benefit/cost).

$$COP \equiv \frac{Q_E}{Q_G} = \frac{Heat\ removed\ in\ Evaporator}{Heat\ input\ in\ Generator} \qquad (12.3.7)$$

If we solve eqs.(13.3.4 & 6) for Q_E and Q_G, we can obtain the ratio

$$COP \equiv \frac{Q_E}{Q_G} = \frac{(1 - y)[h_3^L - h_4]\dot{m}}{(\dot{m} + \dot{M})h_9 - \dot{M}h_{10} - \dot{m}h_1} = \frac{R_m(1 - y)[h_3^L - h_4]}{h_9 - (1 - R_m)h_{10} - R_m h_1} \qquad (12.3.8)$$

where R_m is the mass flow ratio, i.e. $R_m \equiv m/(m+M)$ between stream 1 and stream 10. Once we have a thermodynamic model for the working fluid, we can evaluate the enthalpies in (12.3.8) and thus the COP. By changing the physical parameters of the working fluids, we can optimize the operating conditions as well as the types of working fluids that will give higher COP.

12.4. The Thermodynamic Formulas for Enthalpy Calculation

We know from thermodynamic principles that the enthalpy is related to the Gibbs free energy by the *Gibbs-Helmholtz relation*

$$\left(\frac{\partial(G/T)}{\partial(1/T)}\right)_P = H, \quad or \quad \left(\frac{\partial(G/T)}{\partial(T)}\right)_P = -\frac{H}{T^2} \qquad (12.4.1)$$

This will be the basis of the following developments. We adopt the particular formulation due to Silvester and Pitzer.[84] Let the electrolyte solution be formed from one solvent a (water) and one salt 2 (LiBr). The total enthalpy H of this solution can be expressed as the sum of the enthalpy H_2^{∞} of the solution at infinite dilution (i.e. when mole fraction $x_2 = 0$) and the relative enthalpy Q_R. (Notation: superscript 0 means that the quantity is evaluated at the "pure" state, and ∞ means that the quantity is evaluated at infinite dilution at $x_2 = 0$). Q_R. is defined by (12.4.2).

$$H \equiv H_2^{\infty} + Q_R \qquad (12.4.2)$$

We calculate H_2^{∞} and Q_R separately. First we examine Q_R. Since the reference state in (12.4.2) is at infinite dilution, and the reference for the ion activity coefficients is also at infinite dilution, there is a match. The enthalpy Q_R is obtained simply by applying (12.4.1) to the excess Gibbs free energy G^{ex} of the electrolyte solution (excess over the infinite dilution state)

$$\left(\frac{\partial(G^{ex}/T)}{\partial T}\right)_{P,m} = -\frac{Q_R}{T^2} \qquad (12.4.3)$$

We know that the excess Gibbs free energy G^{ex} is the difference between the total solution G and the infinite dilution state G^{∞}.

$$G^{ex} = G - G^{\infty} = n_a \overline{G}_a + n_2 \overline{G}_2, \qquad where$$

$$\overline{G}_a = \frac{n_2}{n_a} \nu RT (1 - \phi_m),$$

$$\overline{G}_2 = \nu RT \ln \gamma_{\pm}$$

$$G^{ex} = \nu n_2 RT (1 - \phi_m + \ln \gamma_{\pm})$$

(12.4.4)

where we have applied the partial molar quantities (the over bar $^{-}$ means partial molar quantity), and used their connections to the osmotic coefficient ϕ_m (on molality scale) and to the mean activity coefficient $\gamma_{\pm}$. We shall use the molality scale throughout here. Thus $n_2 = m_2$ (the molality of salt 2). $\nu = \nu_+ + \nu_-$, the sum of stoichiometric coefficients.

Note that the solution is at temperature T, pressure P, and composition m_2 (molality of salt 2). At infinite dilution, the salt concentration m_2 goes to zero. Based on the properties of the partial molar enthalpies, the total enthalpy at infinite dilution is decomposed to

$$H_2^{\infty} = n_a \overline{H}_a^{\infty} + n_2 \overline{H}_2^{\infty} = n_a \underline{H}_a^0 + n_2 \overline{H}_2^{\infty}$$

(12.4.5)

As $m_2 \rightarrow 0$, the partial molar enthalpy $\overline{H}_a^{\infty}$ of the solvent a is the pure solvent enthalpy $\overline{H}_a^{\infty} = \underline{H}_a^0$ at same T and P. (Underscore __ means pure state specific property). The pure solvent enthalpy is to be obtained from handbooks or data banks (for water, from the steam tables). The infinitely dilute salt partial molar enthalpy $\overline{H}_2^{\infty}$ is obtained from the following procedure. In statistical mechanics we know that the chemical potential μ_i/kT is expressed as

$$\beta \mu_i = \ln(\rho_i \Lambda_i^3) + \beta \mu_i^{ex} = \ln\left(f_i \frac{\Lambda_i^3}{kT} \right) = \ln\left(\frac{\Lambda_i^3}{kT} \right) + \ln(K_i x_i \gamma_i^{\infty})$$

(12.4.6)

where the excess chemical potential $\beta \mu_i^{ex}$ (excess over the ideal gas value) can be calculated by the direct chemical potential formula[60]

$$\beta \mu_i^{ex} = \sum_{j=a,2} \rho_j \int dr \left[\ln y_{ji} - h_{ji} - \frac{h_{ji} \gamma_{ji}}{2} + B_{ji} \gamma_{ji} - S_{ji} \right]$$

(12.4.7)

where y, h, γ, B, and S are the correlation functions (the cavity correlation, total correlation, indirect correlation, bridge function, and the star function, respectively). In (12.4.6), we have also shown the classical definition of the fugacity f_i and the activity coefficient γ_i^{∞} in an asymmetric convention, i.e., with reference to the infinite dilute state. K_i is the Henry constant,

$$f_i = K_i x_i \gamma_i^{\infty} \tag{12.4.8}$$

Since $\rho_i = \rho x_i$. We can cancel the common factors in (12.4.6)

$$\ln(\rho kT) + \beta \mu_i^{ex} = \ln(K_i \gamma_i^{\infty}) \tag{12.4.9}$$

At infinite dilution (as $x_i \to 0$, $\gamma_i^{\infty} \to 1$), the solution density ρ becomes the pure solvent density ρ_a. Henry's constant K_i can be related to the molecular formula (12.4.6 & 7) via

$$\ln K_i = \ln(\rho_a kT) + \rho_a \sum_{j=+,-} \int dr \left[\ln y_{ja} - h_{ja} - \frac{h_{ja}\gamma_{ja}}{2} + B_{ja}\gamma_{ja} - S_{ja} \right]^{\infty} \tag{12.4.10}$$

where the superscript ∞ on the brackets indicates that the correlation functions inside are evaluated at infinitely dilution $x_2 \to 0$. As a first approximation, we set all the correlation functions to zero, and retain only the first term $\ln(\rho_a kT)$. Applying the mean values "$\pm$"

$$\begin{aligned}
&\ln(\rho_a kT) = \ln K_i, \\
&\ln K_+ = \ln(\rho_a kT), \qquad \ln K_- = \ln(\rho_a kT), \\
&\nu_+ \ln K_+ + \nu_- \ln K_- \equiv \nu \ln K_{\pm} = \nu \ln(\rho_a kT)
\end{aligned} \tag{12.4.11}$$

where we have used the mean electrostatic convention of $\pm$ to define a Henry's constants $K_{\pm}$.

Note that when m_2 molal of salt dissolves in water, the dissociation gives $\nu_+ m_2$ moles of cations and $\nu_- m_2$ moles of anions with total number of ions = νm_2. For example, the LiBr salt dissolved in water will form the following chemical reaction

$$LiBr \xrightarrow{\text{water}} v_+Li^+ + v_-Br^-, \qquad (12.4.12)$$

$$\beta\mu_2 = v_+\beta\mu_+ + v_-\beta\mu_-$$

We have written down the chemical potential balance at equilibrium. Let us define a mean chemical potential $\beta\mu_\pm$ as

$$\beta\mu_2 = v_+\beta\mu_+ + v_-\beta\mu_- \equiv v\beta\mu_\pm \qquad (12.4.13)$$

Retracing the above developments (12.4.6 & 8 & 9), we have at infinite dilution

$$\beta\mu_2^\infty = v_+\beta\mu_+^\infty + v_-\beta\mu_-^\infty \equiv v\beta\mu_\pm^\infty \qquad (12.4.14)$$

Thus the enthalpy $\overline{H}_2^\infty$ is, from (12.4.1)

$$\left(\frac{\partial\beta\mu_2^\infty}{\partial T}\right)_P = -\frac{\overline{H}_2^\infty}{T^2} = v\frac{\partial\ln K_\pm}{\partial T} - \frac{v}{T} \qquad (12.4.15)$$

Combining eqs.(12.4.3 & 4 & 13), finally the total enthalpy of the electrolyte solution[36] is

$$H = H^\infty + Q_R = vm_2RT^2\frac{\partial}{\partial T}\left[\phi_m - \ln\gamma_\pm - \ln K_\pm\right]_{P,m_2} + vm_2RT + n_a\underline{H}_a^0 \qquad (12.4.16)$$

The quantities: the osmotic coefficient and the mean activity coefficient for the electrolyte solutions were given previous in Chapter 7 on the MSA theory. These can be used here. With (12.4.16), we have a methodology at hand for calculating the enthalpies of electrolyte solutions in the operating units of the AbC and finally for the entire refrigeration cycle. Although we have derived the formulas using the mole fractions, the same derivations could have been made with molality. The final equation (12.4.16) here is expressed in molality units for future use.

12.5. The Efficiencies and Enthalpies in Absorption Cycle

The candidates for new working fluids are of three types:[36]

(1) <u>Aqueous electrolyte solutions</u>: with salts LiBr, LiCl, LiClO$_4$, KSCN, and LiCl+ LiNO$_3$.

(2) <u>Aqueous + cosolvent electrolyte solutions</u>: LiBr+ ethylene glycol + water, LiCl+ methanol + water, LiBr + ammonia + water.

(3) <u>Ammoniac electrolyte solutions</u>: Sodium thiocyanate NaSCN+ ammonia, LiNO$_3$ + ammonia, and NaSCN + LiNO$_3$ + ammonia.

12.5.1. Enthalpies and vapor pressures of electrolyte solutions

We show some examples of the enthalpy calculations below. For the aqueous LiBr system, the enthalpies at different temperatures are calculated[37,38] with the above method and compared with experimental data in Figure 12.5.1. Similarly, an enthalpy calculation was made for a mixed solvent system: Water + ethylene glycol + LiBr (Figure 12.5.2). The calculated results give correct trends, with some deterioration only at high temperatures.

The vapor pressures of water for the LiBr-water-ethylene glycol system[37,38] are shown in Figure 12.5.3. Those of the NaSCN-ammonia system[37,38] are shown in Figure 12.5.4. Other salt systems show similar agreement.[36]

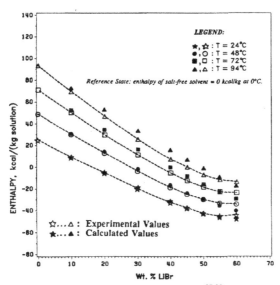

Figure 12.5.1. Enthalpy-concentration diagram[37,38] *of aqueous LiBr solutions at temperatures 24, 48, 72, and 94⁰C. Empty symbols: data from Ellington. Filled symbols: calculated from the MSA method. The dashed lines go through the data to guide the eye.*

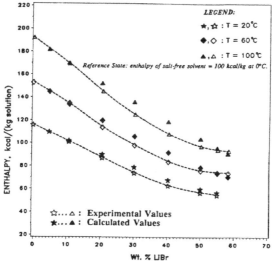

Figure 12.5.2. Enthalpy-concentration diagram[37,38] *of LiBr in water + ethylene glycol binary solvents. The weight ratio of ethylene glycol to water is 0.3445:1. The temperatures are 20, 60, and 100⁰C. Empty symbols: data from Iyoki and Uemura. Filled symbols: calculated from the MSA method. The dashed lines go through the data to guide the eye.*

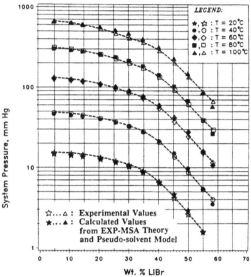

Figure 12.5.3. The vapor pressures of the system LiBr in water+ethylene glycol binary solvents. The temperatures range from 20°C to 100°C. . The weight ratio of ethylene glycol to water is 0.3445:1. Empty symbols: experimental data from Iyoki and Uemura. Filled symbol: from a modified MSA method.[36]

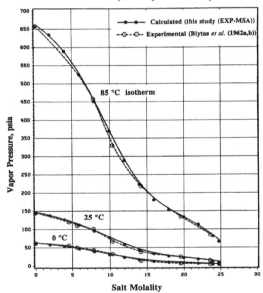

Figure 12.5.4. The vapor pressures of the system: NaSCN in ammonia solvent. The temperatures range from 0°C to 85°C. Empty symbols: experimental data from Blytas et al. Filled symbol: from a modified MSA method.[36]

12.5.2. The coefficients of performance (COP)

With enthalpy calculated, we can evaluate the COP[36] according to eq.(12.3.8). Figure (12.5.5) shows the COP as a function of the generator temperature T_G for three systems: (i) Aqueous LiBr system; (ii) LiBr+ water +ethylene glycol system; and (iii) NaSCN + ammonia system.

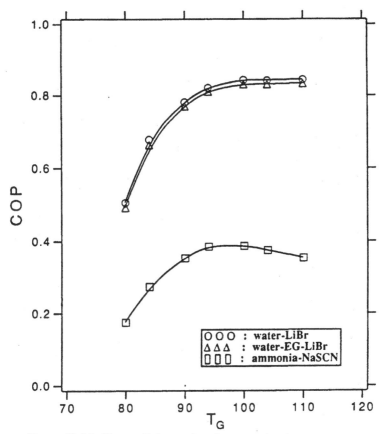

Figure 12.5.5. The coefficients of performance for three systems: (i) aqueous LiBr; (ii) Aqueous LiBr with cosolvent ethylene glycol; and (iii) NaSCN in Ammonia. T_G is the generator temperature in degrees C.

It is seen that the COP increases with generator temperature (the hotter in the generator, the more efficient is the cycle COP). Systems i and ii have similar COP (from 50% to 80%). The ammoniac system has lower COP (from 20% to 40%). This is due essentially to the different latent heats of vaporization of water (with high latent heat value) versus ammonia (with low latent heat value).

The COP calculations can also be made for other types of salts such as $LiNO_3$ and $LiCl$ with cations and anions of different sizes. Also different valence types can be investigated. The MSA method enables us to "screen" candidates. Since large number of salts may qualify for use as alternative working fluids, the analytical method (MSA) offers a fast method of screening the salts with regard to their efficiency and performance. The salts with desirable characteristics can be selected for actual application.   Such an evaluation has been carried out.[36] A summary is given below:

(1) Cations and anions with smaller radii improve the COP by decreasing the circulation ratio $R_m = m/(n+M)$.

(2) Systems with 1-1 type electrolytes give higher COP than the 1-2 and 2-1 electrolytes because they have lower heat capacities at the same operating conditions.

(3) Salts with larger molecular weights give better COP owing to lower circulation ratios.

(4) Adding the cosolvent decreases the circulation ratio significantly and improves COP.

(5) Variation of the dielectric constants of the solvents has little effect on the COP.

The above rules of thumb may occasionally give conflicting demands since it is a parametric study. For example, in actual salts, ions with small radii necessarily have smaller molecular weights. So it is difficult to satisfy both conditions (1) and (3). However, these rules help in identifying alternative salt solutions. The fact LiBr is a popular choice is because it satisfies a number of the criteria above. In addition, other economical and environmental factors such as solubility, corrosion, toxicity, environmental suitability, cost, chemical stability, and availability should all be considered in the final choice.

Exercises:

12.1. Using the water-LiBr binary refrigerant/absorbent pair, find the coefficient of performance (COP) for the following operating conditions: temperature of condenser $T_C = 104°F$, evaporator $T_E = 50°F$, absorber $T_A = 104°F$, Heat exchanger (HExg) temperature approach = 36°F, and generator $T_G = 100°F$. Refer to Figure 12.1.1 (b). Use the formula (12.3.8).

12.2. Using the ammonia-NaSCN binary refrigerant/absorbent pair, find the coefficient of performance (COP) for the following operating conditions: temperature of condenser $T_C = 86°F$, evaporator $T_E = 50°F$, absorber $T_A = 86°F$, Heat exchanger (HExg) temperature approach = 36°F, and generator $T_G = 100°F$. Refer to Figure 12.1.1 (b). Use the formula (12.3.8) for COP.

12.3. Using the water-ethylene glycol-LiBr ternary refrigerant/absorbent /cosolvent mixture, find the coefficient of performance (COP) for the following operating conditions: temperature of condenser $T_C = 104°F$, evaporator $T_E = 50°F$, absorber $T_A = 104°F$, Heat exchanger (HExg) temperature approach = 36°F, and generator $T_G = 100°F$. Refer to Figure 12.1.1 (b). Use the formula (12.3.8).

Chapter 13

Application:
Amine Solutions in Acid Gas Treating

13.1. Introduction

Natural gas produced from the well heads contains many "impurities" and "diluents" such as nitrogen, helium, and organic compounds that need be removed before being used as fuel. Among these are the hydrogen sulfide and carbon dioxide which are called acid gases. The removal of these gases is called *acid gas treating*, or equivalently *natural gas processing*, *gas sweetening*, or *gas conditioning*. It is a major operation in natural gas processing. One class of the solvents used by industry in treating is the aqueous solutions of various amines— such as *monoethanolamine* (MEA), *diethanolamine* (DEA), and *N-methyl-diethanolamine* (MDEA). Such amines will ionize in aqueous environments into onium ions (positively charged ions). In addition, the hydrogen sulfide and carbon dioxide when dissolved in water will also dissociate into ions (negatively charged ions). The anions are neutralized by the cations. There are quite a number of dissociation reactions and chemical reactions taking place. The equipment used to remove the acid gases is an absorption tower. Countercurrent contact of the "sour" natural gas with the liquid amine solution takes place in the tower. The acid gases are neutralized by the aqueous amines and subsequently removed from the gas stream. For more detailed description of industrial practices and historical perspectives, the reader is referred to Astarita[2] (1983) and Kohl and Riesenfeld[54] (1985).

The solution thermodynamics for acid gas treating is complicated by the existence of ionic species and many of the chemical reactions. The major quantities to be controlled are the *vapor pressures* of H_2S and CO_2 in the outgoing "sweetened" gas stream in relation to the amounts (concentrations) of acid gases absorbed in the amine solutions (i.e., the *loading*). Acid gas in the gas phase has a vapor pressure, say in kPa, that is contolled. The amount of acid gas that the amine can carry in *moles of acid gas per mole of amine* is the loading L. The environmental limit on H_2S in vapor phase is less than 4 ppm. The "slip" of CO_2 (amount of CO_2 after treating in the gas stream) can be high (5%-15%), depending on the required heating value of the natural gas. To obtain the partial pressures of the acid gases in the vapor phase during design, we need to carry out vapor-liquid equilibrium calculations. The treating of acid gas necessarily involves the entire discipline of chemical engineering: solution thermodynamics, chemical equilibria, phase equilibria, mass transfer, and the enhancement factor (the extra amount or increased rate of absorption of the acid gases into the liquid phase facilitated by the chemical reactions). We shall concentrate on the application of the electrolyte theory to the acid gas/aqueous amine solution thermodynamics and to the vapor-liquid equilibrium in this chapter.

13.2. Overview of Acid Gas Treating

13.2.1. The absorption process

Acid gas removal is carried out in a conventional absorption tower with countercurrent contact of the gas stream with the scrubbing liquid. The schematics of absorption towers are shown in Figures 13.2.1.

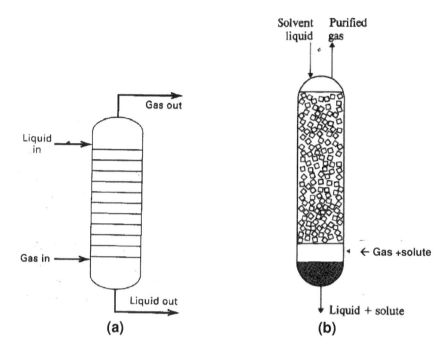

Figure 13.2.1. Absorption towers with countercurrent flows of gas stream (bottom up) and liquid stream (top down). (a) Tray tower. The "ladder rungs" symbolize the stages or trays to promote the liquid-vapor contact (mixing) so that mass transfer of acid gases can take place. (b) Packed tower where random packings are used for mixing the vapor and liquid streams.

Absorption towers use a number of contacting trays where gas and liquid can mix to promote mass transfer (the gas rising up would bubble through a layer of liquid on the tray.) Modern design replaces the trays with random packings (thousands of small cut tubes or doodads of twisted shapes and sizes, ranging from a fraction of an inch to few inches). The liquid would wet the surfaces of the packings, and gas would pass near by intimately contacting the wetted surfaces. Mass transfer takes place between the liquid film and the gas phase. If we magnify the interface between the liquid and the vapor, we would see the mass transfer and chemical reactions as shown in Figure 13.2.2. The acid gas-laden amine stream exits the bottom of the absorber tower and is pumped to a regeneration unit (heater) to recover the amines (which are costly) and recycled. A simplified flow chart of acid gas treating is shown in Figure 13.2.3.

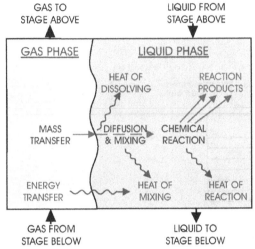

Figure 13.2.2. The gas-liquid interface (the vertical wavy line) where mass transfer (of acid gases) takes place from the gas into the liquid. There would be heats of dissolution, of mixing, and of chemical reactions. The diffusion rate is enhanced due to the chemical reactions that deplete the acid gas concentrations in the bulk liquid. The ratio to the non-reacting case is called the enhancement factor.

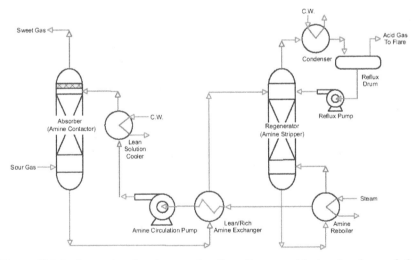

Figure 13.2.3. Conventional amine treating flow diagram with the absorber and the regenerator. The sour gas (natural gas containing CO_2 and H_2S) is "cleansed" or "sweetened" in the absorber. The loaded amine solution is circulated to the regenerator, purified and recycled.

13.2.2. Amine solutions and the chemical reactions

Industry uses amino-alcohols because they have a hydrophilic moiety (the hydroxyl groups, for the sake of dissolving in water) and a basic moiety (to react and neutralize the acid gases). Commonly used amines are MEA (mono-ethanolamine), DEA (di-ethanoamine), MDEA (N-methyl-di-ethanoamine) and their blends. Other types of amines such as TEA (tri-ethanolamine), DIPA (di-isopropanolamine), and DGA (di-glycolamine, or hydroxyaminoethyl ether), have also found application. To conserve energy, sterically hindered amines have been manufactured and entered the market place. Table 13.2.1 gives the chemical formulas of these amines. According to chemistry, amines are classified into primary, secondary, and tertiary amines depending on how many hydrogen atoms on the nitrogen atom have been substituted by the alkanol groups. The following diagram shows that the mono-ethanolamine (MEA) is a primary amine (only one hydrogen atom on **NH₃** has been substituted by the ethanol group).

Primary
Amino
group

Hydroxyl
group

monoethanolamine

When two hydrogen atoms on **NH₃** have been substituted such as in DEA (diethanolamine), it is a secondary amine. When all three hydrogens on **NH₃** are substituted by ethanol moieties, we have TEA (triethanolamine), a tertiary amine.

Table 13.2.1. Common amines used in acid gas treating.

MEA (Monoethanolamine)	$H_2N-CH_2-CH_2-OH$
DEA (Diethanolamine)	$HN=[CH_2-CH_2-OH]_2$
MDEA (Methyldiethanolamine)	$H_3C-N=[CH_2-CH_2-OH]_2$
TEA (Triethanolamine)	$N\equiv [CH_2-CH_2-OH]_3$
DIPA (Diisoprpanolamine)	$HN=[CH_2-(HCCH_3)-OH]_2$
DGA (Diglycolamine)	$H_2N-CH_2CH_2-O-CH_2CH_2-OH$

MEA will receive a proton in water and become protonated according to the chemical reaction

$$HO(CH_2)(CH_2)NH_2 + H_3O^+ \leftrightarrow HO(CH_2)(CH_2)NH_3^+ + H_2O$$

Similarly, other amines will undergo protonation in aqueous environment. These protonated ammonium-like ions can "neutralize" the anionic groups from acid gases. In this Chapter, we shall study typical amines such as MEA, DEA, and MDEA. Carbon dioxide can react to form carbamates with the primary amines and secondary amines (but not with tertiary amines). We classify the reactions into dissociation, protonation and carbamation reactions. The charges on all anions of course are balanced out by all cations.

Dissociation reactions for water, CO_2, H_2S and amines

Dissociation of water:

$$2\,H_2O \longleftrightarrow H_3O^+ + OH^- \qquad (13.2.1)$$

Formation of bicarbonate and carbonate:

$$CO_2 + 2\,H_2O \longleftrightarrow HCO_3^- + H_3O^+ \qquad (13.2.2)$$
$$HCO_3^- + H_2O \longleftrightarrow CO_3^= + H_3O^+ \qquad (13.2.3)$$

Dissociation of hydrogen sulfide in water:

$$H_2S + H_2O \longleftrightarrow HS^- + H_3O^+ \qquad (13.2.4)$$
$$HS^- + H_2O \longleftrightarrow S^= + H_3O^+ \qquad (13.2.5)$$

Protonation of amines:

(MEA)
$$HO(CH_2)_2\text{-}NH_2 + H_3O^+ \longleftrightarrow HO(CH_2)_2\text{-}NH_3^+ + H_2O \qquad (13.2.6)$$

(DEA)
$$[HO(CH_2)_2N]_2\text{-}NH + H_3O^+ \longleftrightarrow [HO(CH_2)_2N]_2\text{-}NH_2^+ + H_2O \qquad (13.2.7)$$

(MDEA)

$$[HO(CH_2)_2N]_2N\text{-}CH_3 + H_3O^+ \longleftrightarrow [HO(CH_2)_2N]_2NH^+\text{-}CH_3 + H_2O \qquad (13.2.8)$$

Formation of carbamates

The carbamate moiety is ($H_2N\text{-}(C=O)O\text{-}$)

(**R** is any alkyl group). In order for amines to form carbamates with carbon dioxide or bicarbonate, there should be at least one substitutable hydrogen atom left on the nitrogen atom. This rules out the tertiary amines.

(MEA)
$$HO(CH_2)_2\text{-}NH_2 + HCO_3^- \longleftrightarrow HO(CH_2)_2\text{-}NH\text{-}(CO)O^- + H_2O$$
$$(13.2.9)$$
(DEA)
$$[HO(CH_2)_2]_2=NH + HCO_3^- \longleftrightarrow [HO(CH_2)_2]_2=N\text{-}(CO)O^- + H_2O$$
$$(13.2.10)$$

In this study, when all three amines and two acid gases have been dissolved in water, we would have eleven ionic species and six neutral molecular species. (See Table 13.2.2). The ions are assigned sizes (collision diameters). They arise from the bare size of the ions—the Pauling diameters (see Appendix II), plus the hydration shells (shells of water molecules attached to the central ion due to electrostatic attraction). The ion and its coordinated water molecules form a quasi-stable ensemble. Some ions such as lithium Li^+ have a hydration number from 4 to 6 (water molecules). Thus its effective diameter is larger than the bare Pauling crystalline diameter. As the concentration of the salt species increases, there is competition for water molecules—more ions competing for a fixed number of water molecules. Thus the hydration number changes (usually decreases) with increasing concentration of salts. Table 13.2.3 gives the kinetic information on the reaction equilibrium constants for these reactions.

Table13.2.2. Speciation of aqueous amine solutions

Ionic Species	Ion Diameter(Å)	Neutral Species	Group ID (UNIFAC*)
H_3O^+	3	H_2O	H_2O (1)
OH^-	3		
HS^-	(varies)**	H_2S	H_2S (1)
$S^=$	2		
HCO_3^-	(varies)	CO_2	CO_2 (1)
$CO_3^=$	(varies)		
MEAH+	(varies)	MEA	CH_2OH (1),
			(CH_2)-NH_2 (1)
DEAH+	6	DEA	CH_2OH (2),
			$(CH_2)_2$=NH (1)
MDEAH+	(varies)	MDEA	CH_2OH (2),
			$(CH_2)_2$=$NHCH_3$ (1)
MEACOO$^-$	6		
DEACOO$^-$	6		

Only neutral species have UNIFAC group ID. The group types are shown with the number of this type of groups in parentheses.
**The variable ion sizes are to account for hydration diameters in different amine concentrations. Their values are accounted for in the Fortran executable programs attached.*

13.3. The Thermodynamic Framework

Solution thermodynamics is above all the study of activity coefficients and chemical potentials. Historically, several industrial modeling methods have been developed for acid gas treating. The earliest was the Kent-Eisenberg[51] correlation. Later, more sophisticated thermodynamic models have been proposed: for example, the methods of Deshmukh-Mather,[25] Chakravarty-Weiland[18] and coworkers, and Chen and coworkers.[75] We introduce here an activity approach based on the molecular MSA model and embellished with the group contribution (GC) methods. We shall call it the *ElecGC* method.

In acid gas treating for our purposes, more than 10 simultaneous reactions and over 17 distinct species are formed. (See Tables 13.2.2 & 3). We classify the species into ionic and neutral types. The mean spherical approach (MSA) is used for the ionic activity coefficients. For neutral species, we employ the group contribution approach—i.e. the UNIFAC (Universal Functional Activity Coefficient) method[29]. In Chapter 7 we have already discussed in some details the approach of

MSA. Since the ions are considered as charge hard spheres (a hard sphere core with embedded positive or negative charges), we also need to consider the reference system of hard spheres (plus the mixtures of them). Thus we shall first introduce the chemical potentials of the hard spheres. Then we proceed to the primitive model of electrolytes.

Table 13.2.3. Equilibrium K-values for the chemical reactions in amine solutions. (plus Henry's constants for H_2S & CO_2) (Note; T in Kelvin)

(1)	$2H_2O = H_2O^+ + OH^-$
	$\ln(K) = 132.899 - 13445.9/T - 22.4773*\ln(T)$
(2)	$2H_2O + CO_2 = H_3O^+ + HCO_3^-$
	$\ln(K) = 231.465 - 12092.1/T - 36.7816*\ln(T)$
(3)	$H_2O + HCO_3^- = H_3O^+ + CO_3^=$
	$\ln(K) = 216.049 - 12431.7/T - 35.4819*\ln(T)$
(4)	$H_2O + H_2S = H_3O^+ + HS^-$
	$\ln(K) = 214.582 - 12995.4/T - 33.5471*\ln(T)$
(5)	$H_2O + HS^- = = H_3O^+ S^=$
	$\ln(K) = -32 - 3338/T$
(6)	$MDEAH^+ + H_2O = MDEA + H_3O^+$
	$\ln(K) = 9.4165 - 4235.98/T$
(7)	$MEAH^+ + H_2O = MEA + H_3O^+$
	$\ln(K) = 4.907365 - 6166.1156/T - 0.0009848*T$
(8)	$DEAH^+ + H_2O = DEA + H_3O^+$
	$\ln(K) = -13.2964 - 4214.0761/T + 0.0099612*T$
(9)	$MEACOO^- + H_2O = MEA + HCO_3^-$
	$\ln(K) = 0.030669 - 2275.19/T$
(10)	$DEACOO^- + H_2O = DEA + HCO_3^-$
	$\ln(K) = 1.655469 - 2057.4377/T$
(11)	$H_3PO_4 + H_2O = H_2PO_4^- + H_3O^+$
	$\ln(K) = 6.468325 - 1840.479/T - 0.031053*T$
(12)	$H_2PO_4 + H_2O = HPO_4^= + H_3O^+$
	$\ln(K) = 520.8779 - 18080.01989/T - 89.7846*\ln(T) + 0.103053*T$
(13)	$HPO_4^= + H_2O = PO4^{-3} + H_3O^+$
	$\ln(K) = 6075.892 - 168503.164/T - 1059.78875*\ln(T) + 1.663444*T$
(14)	H_2S: Henry's constant
	$\ln(H) = 358.899 - 13236.8/T - 55.0551*\ln(T) + 0.059565*T$
(15)	CO_2: Henry's constant
	$\ln(H) = 170.7126 - 8477.711/T - 21.9574*\ln(T) + 0.005781*T$

13.3.1. Hard sphere mixtures

The only way to distinguish different species of hard spheres is by their sizes (diameters). They have no soft energy of interaction. The interaction energy is expressed as

$$u_{ij}(r) = \infty, \qquad r \leq d_{ij}$$
$$u_{ij}(r) = 0, \qquad r > d_{ij} \qquad\qquad (13.3.1)$$

where d_{ij} is the diameter of the collision diameter between species i and species j, $i,j=1,2,3,\ldots,n$. The cross interaction has diameter $d_{ij} = (d_{ii}+d_{jj})/2$, the arithmetic mean. This is called the "additive diameter" assumption. The structure and thermodynamics of hard spheres have been extensively studied in statistical mechanics.[59,87]

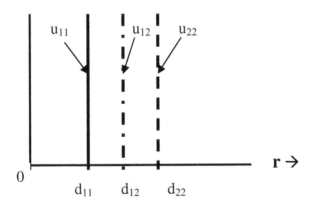

Figure 13.3.1. The hard sphere potentials for a binary mixture of species 1 and 2. d_{ij} (i,j=1,2) are the diameters of the i-j interaction. Since the interactions are infinitely repulsive, the energies are represented by vertical lines.

For thermodynamic properties, one can start with the equation of state and derive therefrom other properties. For mixtures, there are quite a few choices of equations: (i) the Mansoori-Leland-Carnahan-Starling (MLCS) equation[87] used before; or (ii) the Boublik equation.[12] We have used the MLCS equation in Chapter 7. Due to the high accuracy and geometric nature of (ii), we shall use the Boublik equation in this Chapter. (The two equations are equivalent for hard spheres).

Boublik equation of state for hard sphere mixtures

Boublik equation[12] was developed not only for hard spheres but also for hard convex bodies (such as spherocylinders or ellipsoids). The pressure is characterized by the sphere volumes (b_i) occupied by the hard bodies, the surface areas (s_i) of the spheres (available for surface contacts), and the mean radii ($\bar{r}_i$) of curvature (the length scale of the molecules). To be precise, these factors are formally defined for a mixture of hard spheres. Using mole fractions x_i for species i

$$\bar{r} \equiv \sum_i x_i \bar{r}_i$$

$$s \equiv \sum_i x_i s_i = \sum_i x_i (4\pi r_i^2)$$

$$b \equiv \sum_i x_i b_i = \sum_i x_i (\frac{4\pi}{3} r_i^3) \qquad (13.3.2)$$

$$c \equiv \sum_i x_i \bar{r}_i^2$$

where we have also defined a c-factor that is the molar average of the "squares" of the mean radii $\bar{r}_i$ of curvature. (Note for spheres, the mean radius of curvature $\bar{r}_i = r_i$, the radius of the sphere.) These geometric factors appear naturally in the virial expansion of the pressure. Resummation of the virial expansions has inspired the formulation of the **Boublik equation**.[12] In terms of these geometric factors, Boublik[12] proposes

$$\frac{P}{\rho kT} = \frac{1}{1-y} + \frac{3\alpha_1 y}{(1-y)^2} + \frac{3\alpha_2 y^2}{(1-y)^3} - \frac{\alpha_2 y^3}{(1-y)^3} \qquad (13.3.3)$$

where

$$\alpha_1 \equiv \frac{\bar{r}s}{3b}, \qquad \alpha_2 \equiv \frac{cs^2}{9b^2}, \qquad y \equiv \rho b \qquad (13.3.4)$$

y is the packing fraction (just like ζ_3 defined before: the fraction of space physically occupied by the hard spheres—the remainder of the space is of course the free space that is empty of matter). The α-factors are measures of non-sphericity. When the particles are pure spheres, both $\alpha_1 = 1$ and $\alpha_2 = 1$ (Note this is due to the fact that a sphere has $s_i = 4\pi r_i^2$, and $b_i = (4\pi/3)r_i^3$). When the molecules are elongated (as in ethane or naphthalene), the α-factors are greater than 1. After some thermodynamic manipulations (deriving the Helmholtz free energy from the Boublik equation, then differentiating the Helmholtz free energy to obtain the chemical potentials), we obtain the chemical potentials of hard spheres in a mixture as

$$\beta\mu_i^{HSm} - \beta\mu_i^{idgm} = -\ln(1-y)(1-K_i\alpha_2) +$$
$$+ \frac{1}{1-y}\left[(1-\alpha_2)y\frac{b_i}{b} + 3J_i\alpha_1 - K_i\alpha_2\right] + \tag{13.3.5}$$
$$+ \frac{1}{(1-y)^2}\left[(3\alpha_1 - \alpha_2)y\frac{b_i}{b} + K_i\alpha_2\right] - 3J_i\alpha_1$$

where superscripts HSm = hard sphere mixture, and $idgm$ = ideal gas mixture. The values of the hard-sphere chemical potentials are expressed as excesses over the ideal gas counterparts. The excess geometric parameters J_i and K_i are defined by

$$K_i \equiv 2\frac{s_i}{s} + \frac{c_i}{c} - 2\frac{b_i}{b}, \qquad and$$
$$J_i \equiv \frac{r_i}{r} + \frac{s_i}{s} - \frac{b_i}{b}, \tag{13.3.6}$$

We note that this result is equivalent to (7.1.13) and those in literature (see, e.g., Reid and Gubbins[87]). However, the expression (13.3.5) can be used for hard convex bodies as well, and is thus more general.

13.3.2. The Born contribution

When transferring ions from one dielectric medium to another, there is a free energy change, the Born correction

$$\ln \gamma_i^{Born} = \frac{z_i^2 e^2}{2kTd_i}\left[\frac{1}{\varepsilon_{solution}} - \frac{1}{\varepsilon_{water}}\right] \tag{13.3.7}$$

13.3.3. The activity coefficients of ions

We are ready to write down the complete expressions for the activity coefficients of ions (as sum of the MSA contribution, the Born contribution and the hard sphere contribution)

$$\ln \gamma_i^{ion} = \ln \gamma_i^{MSA} + \ln \gamma_i^{Born} + \ln \gamma_i^{HSm}$$

$$\ln \gamma_i^{ion} = -\frac{B\Gamma z_i^2}{1+\Gamma d_i} - \frac{\frac{\pi B}{2\Delta} z_i d_i P_n}{1+\Gamma d_i} - \frac{d_i P_n}{4\Delta}\left[\Gamma a_i + \frac{\pi^2 B}{3\Delta} d_i^2 P\right] - \frac{\pi B z_i}{6}\left[\sum_k \rho_k d_k^2 (\frac{2\Gamma a_k}{4\pi B} + \frac{z_k}{2})\right] +$$

$$+\frac{z_i^2 e^2}{2kTd_i}\left[\frac{1}{\varepsilon_{solution}} - \frac{1}{\varepsilon_{water}}\right] +$$

$$-\ln(1-y)(1-K_i\alpha_2) + \frac{(1-\alpha_2)y\frac{b_i}{b} + 3J_i\alpha_1 - K_i\alpha_2}{1-y} + \frac{(3\alpha_1 - \alpha_2)y\frac{b_i}{b} + K_i\alpha_2}{(1-y)^2} - 3J_i\alpha_1$$

$$(13.3.8)$$

This is the complete expression for the activities of ions in the amine solution. For the special case where $P_n = 0$, the above expression simplifies to

$$\ln \gamma_i^{ion} = -\frac{B\Gamma z_i^2}{1+\Gamma d_i} + \frac{z_i^2 e^2}{2kTd_i}\left[\frac{1}{\varepsilon_{solution}} - \frac{1}{\varepsilon_{water}}\right] +$$

$$-\ln(1-y)(1-K_i\alpha_2) + \frac{(1-\alpha_2)y\frac{b_i}{b} + 3J_i\alpha_1 - K_i\alpha_2}{1-y} + \frac{(3\alpha_1 - \alpha_2)y\frac{b_i}{b} + K_i\alpha_2}{(1-y)^2} - 3J_i\alpha_1$$

$$(13.3.9)$$

For definitions, refer to Chapter 7.

13.3.4. The activity coefficients of neutral species—UNIFAC

For the neutral species such as H_2S, CO_2, water, and the amines (MEA, DEA, and MDEA), we adopt the group contribution correlation UNIFAC. The advantage is that since new amines appear all the time, with the group contribution methods these new amines can be modeled without having to devise new correlations or find new parameters every time they are needed. The UNIFAC method is well-known in literature.[29] We summarize the formulas below for completeness.

Overview

A solution composed of molecules, in UNIFAC, is "decomposed" into an equivalent solution composed of "functional groups" (the chemical moieties that compose the molecules). For example, the solution of ethane molecules (with the moiety the methyl group $-CH_3$) and ethanol molecules (with the moieties the methyl group $-CH_3$, the methylene group $-CH_2$ and the hydroxyl group $-OH$) is replaced by a "soup" of functional groups of methyls $-CH_3$, methylenes $-CH_2$, and hydroxyls - OH. These groups interact and combine to produce the molecular activity coefficients: γ_{ethane} and $\gamma_{ethanol}$. There is a subtle point in choosing the functional groups (e.g. whether $-OH$ is a group, or better yet the larger unit $-CH_2OH$ being a more suitable group). We shall refer to the reference.[112]

UNIFAC activity coefficients of neutral solutes

There are two parts of a UNIFAC activity coefficient: a *combinatorial* (*Comb*) part and a *residual* (*Resid*) part.

$$\ln \gamma_i^{UNIFAC} = \ln \gamma_i^{Comb} + \ln \gamma_i^{Re\,sid} \qquad (13.3.10)$$

The combinatorial part is composed of

$$\ln \gamma_i^{Comb} = \ln \frac{\phi_i}{x_i} + \frac{z}{2} q_i \ln \frac{\theta_i}{\phi_i} + l_i - \frac{\phi_i}{x_i} \sum_k x_k l_k \qquad (13.3.11)$$

where x_i is the mole fraction of the molecular component i. ϕ_i is the volume fraction of the molecular component i; and θ_i is the surface fraction of the molecular component i

$$\phi_i \equiv \frac{x_i r_i}{\sum_k x_k r_k}, \qquad \vartheta_i \equiv \frac{x_i q_i}{\sum_k x_k q_k},$$

$$r_i \equiv \sum_k v_i^k R_k, \qquad q_i \equiv \sum_k v_i^k Q_k, \qquad (13.3.12)$$

where R_k is the volume of a group (moiety) of type k in molecule i. $v_i^k =$ the number of k groups (e.g. methyls) in molecule i (ethane) $=2$. And Q_k is the surface area of group k in molecule i. These values are determined in advance and tabulated for all the moieties (see Table 13.3.1). Also

$$l_i \equiv \frac{z}{2}(r_i - q_i) - (r_i - 1) \tag{13.3.13}$$

where z is the coordination number (number of nearest neighbor molecules surrounding a center molecule. Normally we set $z = 10$). The residual part is given in terms of the exchange energies $a_{jk} = U_{jk} - U_{kk}$:

$$\ln \gamma_i^{\text{Re sid}} \equiv \sum_k v_i^k \left[\ln \Gamma_k - \ln \Gamma_k^i \right] \tag{13.3.14}$$

where Γ_k is the residual contribution from group k to the activity coefficient of molecular species i in solution, while Γ_k^i is the residual contribution from group k to the activity coefficient of i in *pure* liquid i. This subtraction is necessary in order to normalize the molecular activity coefficient of i to unity in pure fluid i.

$$\frac{\ln \Gamma_k}{Q_k} = 1 - \left(\sum_{m=1}^{Ng} \Theta_m e^{-a_{mk}/T} \right) - \sum_{m=1}^{Ng} \frac{\Theta_m e^{-a_{mk}/T}}{\sum_n \Theta_n e^{-a_{nm}/T}} \tag{13.3.15}$$

Note that $\ln \Gamma_k^i$ is obtained when the conditions of pure fluid i is put in place in eq.(13.3.15). The group surface fraction Θ_m and group fraction X_m are similarly defined with the concept of "solution of groups"

$$X_i \equiv \frac{\sum_{j=1}^{Nsp} v_j^i x_j}{\sum_{n=1}^{Ng} \sum_{m=1}^{Nsp} v_m^n x_m}, \qquad \Theta_i \equiv \frac{X_i \Theta_i}{\sum_{n=1}^{Ng} X_n \Theta_n} \tag{13.3.16}$$

Note that N_g is the number of group species (moiety types). In the ethane-ethanol mixture, $N_g = 3$ (namely, the three moieties $-CH_3$, $-CH_2$, and $-OH$). And on the molecular side, N_{sp} is the number of molecular species. For an ethane-ethanol mixture, $N_{sp} = 2$ (i.e., two species ethane and ethanol). The values of R_i, Q_i, and a_{ij} for the groups are listed in the Tables 13.3.1. & 13.3.2.

Table 13.3.1. The Group Volumes R_i and Surface Areas Q_i in UNIFAC

Group Species	Name	R_i	Q_i
H_2O	(water)	0.92	1.4
H_2S	(Hydrogen sulfide)	1.1732	1.07
CO_2	(Carbon dioxide)	1.3	1.12
$-CH_2OH$	(Methylene alcohol)	1.2044	1.124
$-(CH_2)NH_2$	(primary amine)	1.3691	1.236
$=(CH_2)_2NH$	(secondary amine)	1.8813	1.476
$=(CH_2)_2NCH_3$	(tertiary amine)	2.5353	2.02

Activity coefficients of neutral species in an electrolyte solution

We introduced the UNIFAC activity coefficients for the neutral species in a solvent <u>without</u> the ionic species. In case the ionic species are present (such as in an electrolyte solution), the ionic species will cause modifications to the neutral activity coefficients in such a way that the Gibbs-Duhem relation is satisfied. (This is called *thermodynamic consistency* and is an important condition when considering electrolyte solutions!)

The complete expression for neutral species (e.g., H_2S or CO_2) in electrolyte solutions is (via the Gibbs-Duhem equation)

$$\ln \gamma_i^{neutral} \equiv \ln \gamma_i^{UNIFAC} + \frac{\Gamma^3}{3\pi \sum\limits_c \rho_c^0} + \frac{\alpha^2}{8\sum\limits_c \rho_c^0} \left[\frac{P_n}{\Delta}\right]^2 - \frac{\sum\limits_{ion}^{Nion} \rho_{ion}}{\sum\limits_c \rho_c^0}(Z_{ion}^{HS} - 1) \quad (13.3.17)$$

where N_c is the number of species of all neutral molecules (such as H_2S, CO_2, MEA, H_2O, etc.), but excluding any ions. N_{ion} is the number of ionic species (such as OH^-, HCO_3^-, $MDEA^+$, $MEACOO^-$ etc.), while excluding any neutral molecules. The density ρ_c is the number density of the neutral species c. ρ_{ion} is the number density of all ions of species $_{ion}$ in solution. In (13.3.17), $\ln \gamma^{UNIFAC}$ is the expression (13.3.10) defined solely for the neutral species.

Table 13.3.2. The Exchange Energies a_{ij} in UNIFAC Group ID:

1. H_2O	2. H_2S	3. CO_2	4. $-CH_2OH$	5. $-(CH_2)NH_2$	6. $=(CH_2)_2NH$	7. $=(CH_2)_2NCH_3$

(Note: TK = temperature in Kelvin)

$a(1,2)=595.962$
$a(1,3)=269.1645$
$a(1,4)=-83.88$
$a(1,5)=-993.39+252876.78/TK$
$a(1,6)=-168.08+14528.78/TK$
$a(1,7)=58.0$
$a(2,1)=514.797$
$a(2,3)=-463.98+195020.0/TK$
$a(2,4)=700.0$
$a(2,5)=700.0$
$a(2,6)=700.0$
$a(2,7)=700.0$
$a(3,1)=491.145$
$a(3,2)=204.0-55019.8/TK$
$a(3,4)=700.0$
$a(3,5)=700.0$
$a(3,6)=700.0$
$a(3,7)=700.0$
$a(4,1)=93.97$
$a(4,2)=700.0$
$a(4,3)=700.0$
$a(4,5)=-2621.8+890152.94/TK$
$a(4,6)=1416.73-203598.0/TK$
$a(4,7)=352.97-135875.0/TK$
$a(5,1)=715.92-311711.19/TK$
$a(5,2)=700.0$
$a(5,3)=700.0$
$a(5,4)=2777.88-989632.13/TK$
$a(6,1)=-631.0+88626.0/TK$
$a(6,2)=700.0$
$a(6,3)=700.0$
$a(6,4)=-1676.49+402677.0/TK$
$a(7,1)=6.985-78637.4/TK$
$a(7,2)=700.0$
$a(7,3)=700.0$
$a(7,4)=-263.918+168732.6/TK$

The α and Z^{HS}_{ion} are

$$\alpha^2 \equiv \frac{4\pi e^2}{\varepsilon_m kT}, \qquad Z^{HS}_{ion} - 1 \equiv \frac{y}{1-y} + \frac{3\alpha_1 y}{(1-y)^2} + \frac{3\alpha_2 y^2}{(1-y)^3} - \frac{\alpha_2 y^3}{(1-y)^3},$$

$$y = \sum_{ion}^{N_{ion}} \rho_{ion} b_{ion}$$

(13.3.18)

The nonsphericity factors α_1 and α_2 were defined earlier in eqs. (13.3.4). The subscript $_{ion}$ counts only the ionic species, excluding the neutral species. ε_m is the solution permittivity.

13.4. Practical Calculations for Acid Gas Vapor Pressures

The above thermodynamic method has been given the name of *ElecGC* (electrolyte-group-contribution method). It can be applied to the *loading vs. vapor pressure* calculation in acid gas treating. It has been assumed here that thermodynamic vapor-liquid equilibrium has been attained. (Namely not for transient processes or nonequilibrium contact processes). The amine types included in *ElecGC* are MEA, DEA, MDEA, and their blends. The acid gases are H_2S and CO_2. The method can be used for both *LtoP* (given loading L, find the vapor pressures of the acid gases), or *PtoL* (given vapor pressures, find the loading L of amines—the moles of acid gas per mole of amine used). The vapor pressures of acid gases in the sweetened gas stream are indicators of the purity (the smaller the vapor pressures, the cleaner the sweetened gas). *ElecGC* can also calculate the heats of solution, thus it can be use for heat duty calculations. A unique feature is that it furnishes data on speciation (the concentrations of individual species, ionic or neutral, in the solution at any treating condition). The speciation information is critical for corrosion management, enhancement factor estimation, the chemical reaction status, and is a key factor in acid gas treating.

In the vapor-liquid equilibrium (VLE) calculations, one also needs a thermodynamic model for the vapor phase. The vapor phase shall contain, more or less, water vapor, H_2S gas, CO_2 gas, plus the inert air (nitrogen and oxygen). Due to the high boiling points of the amines, there are little or only trace amounts of amines as vapor in the industrial settings. We found that the Peng-Robinson equation of state is sufficient for the vapor phase fugacity calculations. The procedure of VLE calculations is shown in Figure 13.4.1. (i) For the liquid phase, we are given the temperature, the amine strength (wt% of amine in water), the desired loading (LtoP). We want to calculate the vapor pressures of H_2S and CO_2 in the vapor phase. We first use the *ElecGC* method outlined above to calculate the liquid activity coefficients γ_i^L of the ionic components as well as the neutral species therein. (ii) We use the Peng-Robinson equation of state to calculate the vapor phase fugacity coefficients ϕ_i^V of water, H_2S and CO_2. (The amines are assumed to be nonvolatile at the system temperature, since we rarely go beyond 120°C). A bubble point calculation is made using the liquid compositions of water, H_2S and CO_2 from (i). We match the fugacities of water, H_2S and CO_2 between the vapor phase and the liquids phase. We note that the reference state for the liquid activities is the Henry's constants K_i for H_2S and CO_2. Henry's constants K_i are obtained from literature and listed in Table 13.2.3. We show an example below.

Example: Treating of H_2S with 20 wt% aqueous MEA solution

[Objective]: Given the loading L of H_2S in MEA solution, find: the vapor pressure of H_2S in the sweetened gas stream (LtoP Calculation).

<u>System Input Data:</u>
Temperature of system: 40°C
Amine solution: 20 wt% of MEA in water
Loading (moles of H_2S per mole of MEA): $L = 0.00623$ → 1.40

$$\gamma_i \, x_i \, H_i \; = \; \phi_i \, y_i \, P$$

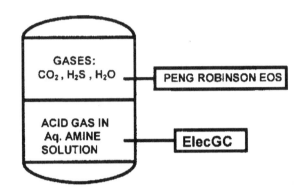

Figure 13.4.1. Procedure of calculation for the vapor-liquid equilibria between the liquid phase and the vapor phase. The fugacties of neutral components in liquid $f_i^L = \gamma_i^L \, x_i K_i$ are matched with those in the vapor phase $f_i^V = \varphi_i^L \, y_i P$.

Solution:

Using the *ElecGC* program attached, we can obtain the outputs on vapor pressure of H_2S and other information on activity coefficients, ion sizes, and speciation as follows:

```
       OUTPUT FROM FORTRAN PROGRAM: ElecGC_Fortran.exe

=================================================================
 Temp=   40.0000000000000      C; or        103.999998092651      F
 [MDEA] =  0.000000000000000E+000M; or  0.000000000000000E+000wt%
 [DEA]  =  0.000000000000000E+000M; or  0.000000000000000E+000wt%
 [MEA]  =  4.09299267363305      M; or  20.0000000000000      wt%
 [H3PO4]=  0.000000000000000E+000 wt%
 Loading: (mole H2S)/(mole tot amine)
    H2S     PH2S(kPa) PH2S(kPa)      Error%  d(A,HS-)  d(A,MEA+)
   Loading   Exp.      Calc.
-----------------------------------------------------------------
   0.00623  0.2900E-02  0.1778E-02   -38.70   11.7762    15.87
   0.18600  0.4490      0.4375        -2.56    3.8695     3.657
   1.16000  777.0       749.4         -3.55    2.5913     2.233
   1.40000  1949.       1620.        -16.88    2.5931     2.235
-----------------------------------------------------------------
     Avg. abs. err. PH2S :   0.211418142892709
```

□

The highlighted numbers are the important quantities thus obtained. The calculated answer is compared with the experimental vapor pressures. The errors range from 2% to 40%. For acid gases, we normally use the logarithmic scale (*lnP* vs. *lnL*). Other data on activity coefficients, ion sizes, and speciation are in a number of other output files.

13.5. Results of Amine Based Acid Gas Treating

The ElecGC method has been applied to acid gas conditions from low loading (*L*=0.003) to high loading (*L*=1.8); and from low vapor pressures (0.0016 kPa) to high pressures (10.8 MPa). Temperature ranges from 40, 70, 100, to 120°C. This range encompasses both the absorber conditions and the regenerator conditions. The amines studied include the MEA, DEA, MDEA, and DEA-MDEA blends. We have also looked at the effects of addition of phosphoric acid on the loading-pressure curve. The conditions that have been tested are listed in Table 13.5.1. The data have been measure by D.B. Robinson, Ltd. We shall examine some typical systems for illustration below: H_2S absorbed in MEA, DEA, and MDEA, also CO_2 absorbed in MEA, DEA, MDEA, and DEA+MDEA blends.

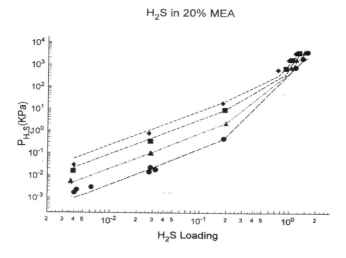

Figure 13.5.1. Loading vs. vapor pressure curves for H_2S absorption in 20 wt% MEA in aqueous solution. The symbols (circles, triangles, squares, and diamonds) are experimental data at temperatures from 40°, 70°, 100°, to 120^oC, respectively. The lines are from ElecGC.

13.5.1. Acid gas loading curves

Figure 13.5.1 gives the vapor pressure of H_2S (from 0.00164 kPa to 3838 kPa) as a function of loading (L = moles of H_2S per mole of MEA used, from 0.0036 to 1.58). Four temperatures are plotted, from bottom up $40°$, $70°$, $100°$, to $120°C$. This is a log-log plot. It is seen that at loading $L<0.2$, the log-log line is almost straight. Beyond $L>0.2$, there is a drastic increase of pressure. The high loading regions represent the end of chemical absorption, and beginning of "physical absorption" where the dissolution of acid gas into the solution is effected by high pressures (i.e. by "squeezing" the gas into the liquid phase), as the amines are being exhausted in the liquid and are no longer able to react with the acid gases.

Table 13.5.1. Acid gas systems examined. (Data from D.B. Robinson)
===

System	Loadings	T $°C$/P kPa
H_2S/MEA 20 wt%	0.0036-1.58	40,70,100,120/0.00164-3838
H_2S/MEA 30 wt%	0.0036-0.20	40,70,100,120/0.00319-32.6
H_2S/MEA 30 wt%	0.0039-1.83	40,70,100,120/0.00513-10823
H_2S/DEA 50 wt%	0.0030-0.18	40,70,100,120/0.00389-51
H_2S/MDEA 23.1 wt%	0.0030-0.20	40,70,100,120/0.00330-35
H_2S/MDEA 50 wt%	0.0030-1.74	40,70,100,120/0.00740-3673
H_2S/MDEA 35 wt%+ DEA 5 wt%	0.0030-1.66	40,70,100,120/0.00476-3679
H_2S/MDEA 35 wt%+ DEA 10 wt%	0.0043-1.62	40,70,100,120/0.00476-4101
H_2S/MDEA 35 wt%+ Phosphoric acid 1 wt%		
	0.0019-1.66	40,70,100,120/0.00529-3545
H_2S/MDEA 35 wt%+ Phosphoric acid 2 wt%		
	0.08-1.54	40,70,100,120/1.55-3559
H_2S/MDEA 35 wt%+ Phosphoric acid 5 wt%		
	0.00335-0.21	40,70,100,120/0.0337-0.21
CO_2/MEA 20 wt%	0.112-1.18	40,70,100,120/0.033-6293
CO_2/MEA 30 wt%	0.03-0.236	40,70,100,120/0.134-14
CO_2/DEA 30 wt%	0.0172-1.22	40,70,100,120/0.169-6353
CO_2/MDEA 23 wt%	0.00334-1.34	40,70,100,120/0.002-5265
CO_2/MDEA 50 wt%	0.00257-1.16	40,70,100,120/0.0122-5327
CO_2/MDEA 35 wt% + DEA 5 wt%	0.00353-1.17	40,70,100,120/0.02-5120
CO_2/MDEA 35 wt% + DEA 10 wt%	0.003-1.16	40,70,100,120/0.01-5355
CO_2/MDEA 35 wt%+ Phosphoric acid 1 wt%		
	0.005-1.17	40,70,100,120/0.05-5318
CO_2/MDEA 35 wt%+ Phosphoric acid 5 wt%		
	0.001-0.93	40,70,100,120/0.04-5297
CO_2+H_2S/MDEA 35 wt%	0.02-0.64	40,70,100,120/0.4-772

Figure 13.5.2 gives the loading curve (in terms of the ionic strength) of H_2S in 30 wt% of DEA. We see a distinct change of slope from low ionic strengths to high ionic strengths.

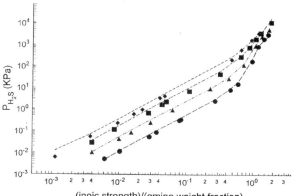

Figure 13.5.2. Loading vs. vapor pressure curves for H_2S absorption in 30 wt% DEA. The symbols (circles, triangles, squares, and diamonds) represent data at temperatures from 40°, 70°, 100°, to 120^oC, respectively.

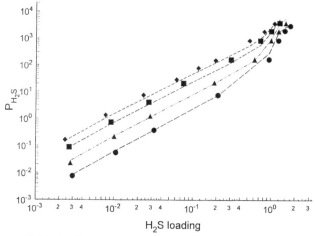

Figure 13.5.3. Loading vs. vapor pressure curves for H_2S absorption in 50 wt% MDEA. The symbols (circles, triangles, squares, and diamonds) represent data at temperatures from 40°, 70°, 100°, to 120^oC, respectively.

Figure 13.5.3 epitomizes the typical behavior of H_2S in 50 wt% MDEA solutions. We note that the points (symbols) are experimental data, and the lines are from the *ElecGC* calculations. All amines in the absorption of H_2S show similar behavior: a linear region on the log-log plot, and a second region of fast rising pressures. Next we show the behavior of CO_2 absorption.

Figure 13.5.4 gives the absorption of CO_2 in 20 wt% MEA. It is of interest to observe that at loading L ~ 0.45, the linear lines begin to bend upward, showing a distinct convex behavior. This is to be understood by looking at the speciation curves (i.e. concentration distributions of all 10 components at varying loading *L*). As will be explained in more details below, the break in the curves is closely related to the formation of carbamates $MEACOO^-$. Since DEA can also form carbamate $DEACOO^-$ with CO_2, the loading curves exhibit similar abrupt changes in slope (at *L*>0.45) (Figure 13.5.6).

We plot the speciation vs. the loading in Figure 13.5.5. We observe the trend lines of individual species as *L* increases. First, we see that the concentration of molecular DEA decreases to form $DEAH^+$ in response to the dissolution CO_2 into HCO_3^-. Next, HCO_3^- reacts with DEA to form the carbamate $DEACOO^-$. As $DEACOO^-$ concentration increases, the HCO_3^- is depleted.

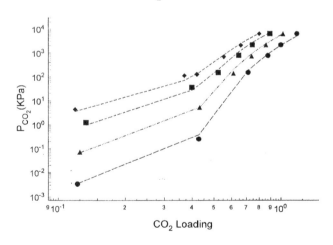

CO$_2$ in 20% MEA

Figure 13.5.4. Loading vs. vapor pressure curves for CO_2 absorption in 20 wt% MEA. The symbols (circles, triangles, squares, and diamonds) represent data at temperatures from 40°, 70°, 100°, to 120°C, respectively.

Speciation for CO_2 in aqueous DEA (30 wt%) at 40° C

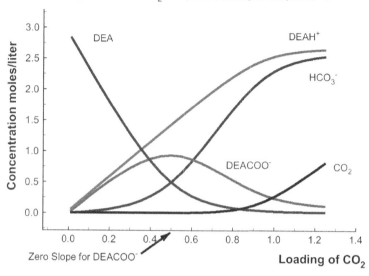

Zero Slope for DEACOO⁻

Loading of CO_2

CO_2 dissolved into 30 wt% DEA at 40°C

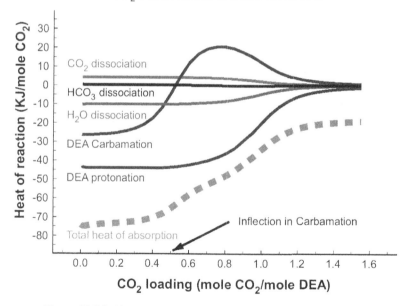

Figure 13.5.5. The Speciation curves and heat of reaction curves for CO_2 in 30 wt% DEA. We observe the formation and dissociation of DEACOO⁻ at L~0.45.

CO$_2$ in 30% DEA solution

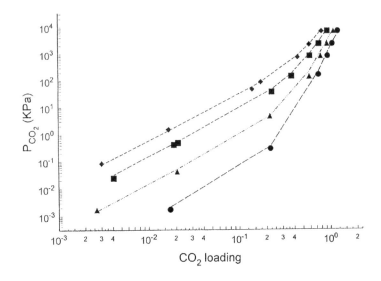

Figure 13.5.6. Loading vs. vapor pressure curves for CO$_2$ absorption in 30 wt% DEA. The symbols (circles, triangles, squares, and diamonds) represent data at temperatures from 40°, 70°, 100°, to 120°C, respectively.

At loading L around 0.45, the CO$_2$ continues to dissolve into the liquid as HCO$_3^-$, thus demanding more DEAH$^+$ in order to balance the increased HCO$_3^-$. Some DEACOO$^-$ starts to dissociate to replenish DEAH$^+$. This point (L~0.45) is pivotal in the explanation of the inflection curves observed in CO$_2$/DEA loading plots (Figs. 13.5.4 & 6). We further notice that the concentration of molecular CO$_2$ is very little in the liquid at the beginning (for L< 0.7), because as soon as CO$_2$ enters the liquid, it reacts to form HCO$_3^-$, which in turn causes the formation of the DEAH$^+$ (to achieve balance of charges). But when DEA molecules are exhausted (used up by DEAH$^+$), i.e. there is no longer any amine left, further absorption must involve neutral CO$_2$ instead of HCO$_3^-$ ions. Thus the molecular CO$_2$ concentration increases after L=7, and pressure increases drastically because the mechanism of dissolution now is due to pressure force (compression), rather than through chemical force. This analysis demonstrates the central role of speciation in understanding acid gas treating. The heats of reaction curves also show the impact of carbamate formation. The total reaction heat generated (dotted line in Figure

13.5.5) is essentially supplied by carbamation, which contributes by far a lion's share of the enthalpy of reaction (see the lower plot of Fig. 13.5.5).

For CO_2 absorption in MDEA solution, there is no carbamate formation (because MDEA is a tertiary amine). Figure 13.5.7 gives the vapor pressure curves. We do not see the breaks in the curves.

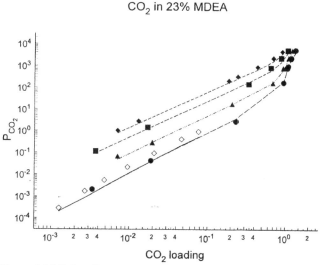

Figure 13.5.7. Loading vs. vapor pressure curves for CO_2 absorption in 23 wt% MDEA. (Legend as before).

It is also of interest to use blended amines (DEA+MDEA) in acid gas treating. MEA and DEA can absorb large amounts of CO_2 because of the formation of carbamates. However, if the emission limits on CO_2 can be relaxed, then we can "slip" the CO_2 during treating by using MDEA, which absorbs H_2S readily, but does not absorb as much CO_2. In addition, carbmation reaction is energy intensive. Thus by reducing DEA, one can save energy. Figure 13.5.8 shows the loading curve using blended amines.

13.5.2. The speciation in amine solutions

The absorption of acid gases is enhanced by the chemical reactions. By tracing the concentration changes of all species, one can tell the degree of absorption. We have shown the CO_2 speciation in DEA above (Fig. 13.5.6). Here (Fig. 13.5 9) is a plot of speciation of CO_2 in aqueous

MDEA (at 50 wt%) and 28°C. It is seen that the MDEAH⁺ ion concentration tracks the HCO3⁻ concentration closely. It shows that the neutralization chemical reactions are primarily responsible for the absorption of acid gases.

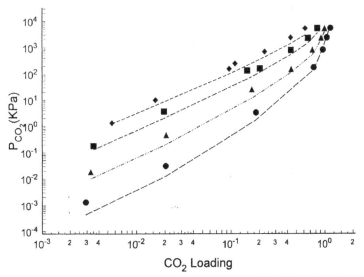

Figure 13.5.8. Loading vs. vapor pressure curves for CO_2 absorption in 35 wt% MDEA and 10 wt% DEA. (Legend as before).

13.5.3. The heat of absorption of acid gases

The amines, upon absorption of acid gases, will release heat. During regeneration, the reverse reactions take place and heat must be supplied. This costs energy. For energy efficiency, one should guard against high enthalpy of absorption. *ElecGC* can estimate the demands for heat in absorption with different amines. The heat of absorption consists of three parts when the acid gas enters the liquid solution: (i) the heat of dissolving (from gas phase to liquid phase), (ii) the heat of mixing (nonideal solution enthalpies), and (iii) heats of reactions (derived from various chemical reactions). A case is shown in Figure 13.5.10 (CO_2 absorbed in 20 wt% MDEA). There is some scatter in the data (filled circles are from the RR-102, Report of GPA). Overall, the prediction by ElecGC (the line) is quite accurate.

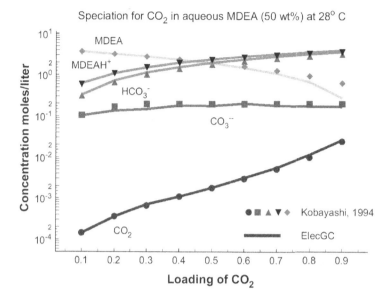

Figure 13.5.9. Speciation curves for CO₂ absorption in 50 wt% MDEA. T=
28°C. Symbols: from measured NMR data. Lines: from ElecGC calculations.

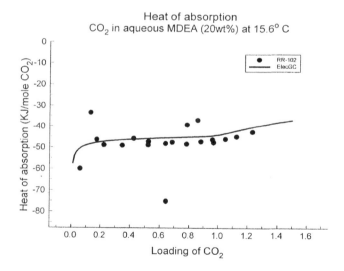

Figure 13.5.10. Heat of absorption of CO₂ absorption in 20 wt% MDEA at T=
15.6°C. Circles: from experimental data (Research report RR-102, GPA).
Line: from ElecGC calculations.

13.5.4. Hydrocarbon solubility in amine solutions

Natural gas is mostly hydrocarbons (methane being the major component). As the raw natural gas stream passes through the amine liquid, it is important to know how much methane and other hydrocarbons can dissolve in the liquid. Few data have been measured and available. Since the solubility of hydrocarbons is very low, the accuracy of measurements is poor and has relatively large errors. *ElecGC* is used to estimate the amount of methane dissolved in 25 wt% DEA with loading of CO_2 at L =0.25 to 0.27. The results are shown in Figure 13.5.11. Symbols are from experimental data. Lines are from *ElecGC*. The prediction by the thermodynamic model is reasonable.

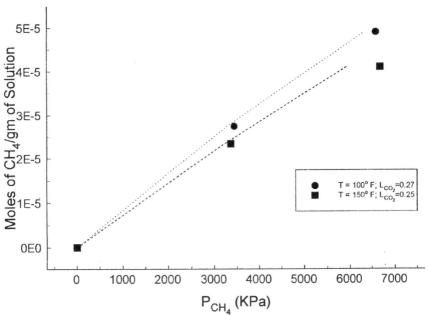

Figure 13.5.11. Methane solubility in aqueous amine solution of DEA loaded with acid gas CO_2. Symbols: from experimental data. Line: from ElecGC calculations

13.6. Remarks on Other Acid Gas Treating Chemicals

We have discussed the common types of amines used in treating acid gases. There are other types of treating solvents. The simplest are the caustic (NaOH) and the potassium carbonate. Sulfolane (tetramethylene sulfone, $C_4H_8O_2S$) solvent is used in removing acid gases and carbonyl sulfide (COS) from raw natural gas streams. Other solvents such as rectisol (methanol-based), selexol (ethylene glycol + dimethyl ethers), and sulfinol (mixtures of tetrahydrothiophene + alkanolamines) are also used. For quite a number of years, sterically hindered amines such as AMP (amino-methyl propanol) and PE (2-piperidine ethanol) have been brought to the market. The Flexsorb series belongs to the hindered amines. Their advantage is low energy costs (low heat duty in regeneration). In addition, N-formylmorpholine is also tested for treating. For these and other compounds, it is possible to use the group contribution methods to model the organic moieties as outline in this Chapter. *ElecGC* is a valid framework for dealing with new amine products. Figure 13.5.12 shows the interconnections between the thermodynamic model *ElecGC* and many of the important tasks in acid gas treating.

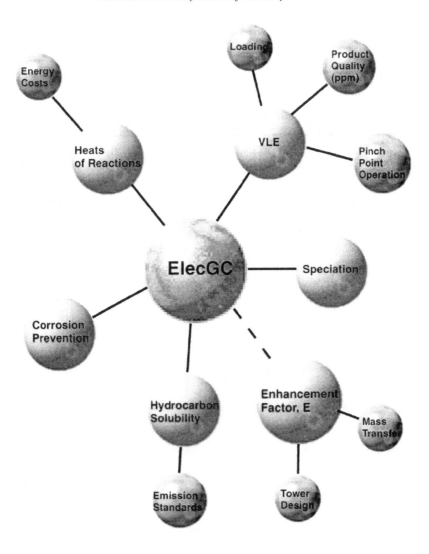

Figure 13.5.12. The connections of the thermodynamic model ElecGC to the operations in acid gas treating. Solid lines: direct connection. Dashed line: indirect connection.

Exercises:

13.1. Use the program "ElecGC" on the CD (or a software*) to find the vapor pressure of H_2S in MEA (20 wt%) solution at 40°C.

13.2. Use the program "ElecGC" on the CD (or a software*) to find the vapor pressure of H_2S in DEA (20 wt%) solution at 40°C.

13.3. Use the program "ElecGC" on the CD (or a software*) to find the vapor pressure of H_2S in MDEA (20 wt%) solution at 40°C. Compare with the above values. Which amine is more effective?

13.4. Use the program "ElecGC" on the CD (or a software*) to find the vapor pressure of CO_2 in MEA (20 wt%) solution at 40°C. Also plot the speciation (concentrations of ions as function of loading).

13.5. Use the program "ElecGC" on the CD (or a software*) to find the vapor pressure of CO_2 in DEA (20 wt%) solution at 40°C. . Also plot the speciation (concentrations of ions as function of loading).

13.6. Use the program "ElecGC" on the CD (or a software*) to find the vapor pressure of CO_2 in MDEA (20 wt%) solution at 40°C. . Also plot the speciation (concentrations of ions as function of loading). Compare with the above values. Which amine is more effective?

13.7. Construct a program of your own (Fortran, C, or other languages), find the vapor pressure of CO_2 in mixed amines MDEA (20 wt%) + DEA (10 wt%) solution at 40°C. You can use the equations developed in this chapter to find the correct vapor pressure value.

13.8. Use the program "ElecGC" on the CD (or a software*) to find the vapor pressure of H_2S in mixed MDEA (20 wt%) and DEA (10 wt%) solution at 100°C. . Also plot the speciation (concentrations of ions as function of loading).

---

*A Windows-interactive (GUI) software for acid gas treating with amine solutions is available for distribution at cost. Contact profllee@yahoo.com for ordering.

APPENDICES

APPENDIX I

Pitzer's Parameters for Electrolyte Solutions

References: (Selected values from the following references)
-- Pitzer, K.S., *Thermodynamics of electrolytes I: Theoretical basis and general equations.* J. Phys. Chem., 77: 268, 1973.
-- Pitzer, K.S., G. Mayorga, *Thermodynamics of electrolytes II: Activity and osmotic coefficients for strong electrolytes with one or both ions univalent.* J. Phys. Chem. 77, 2300 (1973).

Table 1: Inorganic Acids, Bases and Salts of 1-1 Type

Species	β_0	β_1	C^ϕ	Max m	σ
$AgNO_3$	-0.0856	0.0020	0.00591	6.0	0.001
CsBr	0.0279	0.0139	0.00004	5.0	0.002
CsCl	0.0300	0.0558	0.00038	5.0	0.002
CsF	0.1306	0.2570	-0.00430	3.2	0.002
CsI	0.0244	0.0262	-0.00365	3.0	0.001
$CsNO_2$	0.0427	0.0600	-0.00510	6.0	0.004
$CsNO_3$	-0.0758	-0.0669	---	1.4	0.002
CsOH	0.1500	0.3000	---	---	---
HBr	0.1960	0.3564	0.00827	3.0	A
HCl	0.1775	0.2945	0.00080	6.0	A
$HClO_4$	0.1747	0.2931	0.00819	5.5	0.002
HI	0.2362	0.3920	0.00110	3.0	B
HNO_3	0.1119	0.3206	0.00100	3.0	0.001
KBr	0.0569	0.2212	-0.00180	5.5	0.001
$KBrO_3$	-0.1290	0.2565	---	0.5	0.001
KCl	0.04835	0.2122	-0.00084	4.8	0.0005
$KClO_3$	-0.0960	0.2481	---	0.7	0.001
KCNS	0.0416	0.2302	-0.00252	5.0	0.001
KF	0.08089	0.2021	0.00093	2.0	0.001

Table 1 (Continued)

Species	β_0	β_1	C^ϕ	Max m	σ
KH_2AsO_4	-0.0584	0.0626	---	1.2	0.003
KH_2PO_4	-0.0678	-0.1042	---	1.8	0.003
KI	0.0746	0.2517	-0.00414	4.5	0.001
KNO_2	0.0151	0.0150	0.00070	5.0	0.003
KNO_3	-0.0816	0.0494	0.00660	3.8	0.001
KOH	0.1298	0.3200	0.00410	5.5	b
KPF_6	-0.163	-0.282	---	0,5	0.001
LiBr	0.1748	0.2547	0.003	2.5	0.002
LiCl	0.1494	0.3074	0.00359	6	0.001
$LiClO_4$	0.1973	0.3996	0.0008	3.5	0.002
LiI	0.2104	0.373	---	1.4	0.006
$LiNO_2$	0.1336	0.325	-0.0053	6	0.003
$LiNO_3$	0.1420	0.278	-0.00551	6	0.001
LiOH	0.015	0.14	---	4	c
NH_4Br	0.0624	0.1947	-0.00436	2.5	0.001
NH_4Cl	0.0522	0.1918	-0.00301	6	0.001
NH_4ClO_4	-0.0103	-0.0194	---	2	0.004
NH_4NO_3	-0.0154	0.1120	-0.00003	6	0.001
$NaBF_4$	-0.0252	0.1824	0.0021	6	0.006
$NaBO_2$	-0.0526	0.1104	0.0154	4.5	0.004
NaBr	0.0973	0.2791	0.00116	4	0.001
$NaBrO_3$	-0.0205	0.1910	0.0059	2.5	0.001
NaCl	0.0765	0.2664	0.00127	6	0.001
$NaClO_3$	0.0249	0.2455	0.0004	3.5	0.001
$NaClO_4$	0.0554	0.2755	-0.00118	6	0.001
NaCNS	0.1005	0.3582	-0.00303	4	0.001
NaF	0.0215	0.2107	---	1	0.001
NaH_2AsO_4	-0.0442	0.2895	---	1.2	0.001
NaH_2PO_4	-0.0533	0.0396	0.00795	6	0.003

Table 1 (Continued)

Species	$\underline{\beta_0}$	$\underline{\beta_1}$	$\underline{C^\phi}$	Max m	$\underline{\sigma}$
NaI	0.1195	0.3439	0.0018	3.5	0.001
NaNO$_2$	0.0641	0.1015	0.0049	5	0.005
NaNO$_3$	0.0068	0.1783	-0.00072	6	0.001
NaOH	0.0864	0.253	0.0044	6	b
RbBr	0.0396	0.1530	-0.00144	5	0.001
RbCl	0.0441	0.1483	-0.00101	5	0.001
RbF	0.1141	0.2842	-0.0105	3.5	0.002
RbI	0.0397	0.1330	-0.00108	5	0.001
RbNO$_2$	0.0269	-0.1553	-0.00366	5	0.002
RbNO$_3$	-0.0789	-0.0172	0.00529	4.5	0.001
TlClO$_4$	-0.087	-0.023	---	0.5	0.001
TlNO$_3$	-0.105	-0.378	---	0.4	0.001

Table 2: Inorganic Compounds of 2-1 Type

Species	4/3 β_0	4/3 β_1	$2^{5/2}/3\ C^\phi$	Max m	σ
$BaBr_2$	0.4194	2.093	-0.03009	2.0	0.001
$BaCl_2$	0.3504	1.995	-0.03654	1.8	0.001
$Ba(ClO_4)_2$	0.4819	2.101	-0.05894	2.0	0.003
BaI_2	0.5625	2.249	-0.03286	1.8	0.003
$Ba(NO_3)_2$	-0.0430	1.070	---	0.4	0.001
$Ba(OH)_2$	0.2290	1.600	---	0.1	---
$CaBr_2$	0.5088	2.151	-0.00485	2.0	0.002
$CaCl_2$	0.4212	2.152	-0.00064	2.5	0.003
$Ca(ClO_4)_2$	0.6015	2.342	-0.00943	2.0	0.005
CaI_2	0.5839	2.409	-0.00158	2.0	0.001
$Ca(NO_3)_2$	0.2811	1.879	-0.03798	2.0	0.002
$Cd(NO_3)_2$	0.3820	2.224	-0.04836	2.5	0.002
$CoBr_2$	0.5693	2.213	-0.00127	2.0	0.002
$CoCl_2$	0.4857	1.967	-0.02869	3.0	0.004
CoI_2	0.6950	2.230	-0.00880	2.0	0.010
$Co(NO_3)_2$	0.4159	2.254	-0.01436	5.5	0.003
Cs_2SO_4	0.1184	1.481	-0.01131	1.8	0.001
$CuCl_2$	0.3955	1.855	-0.06792	2.0	0.002
$Cu(NO_3)_2$	0.4224	1.907	-0.04136	2.0	0.002
$FeCl_2$	0.4479	2.043	-0.01623	2.0	0.002
K_2CrO_4	0.1011	1.652	-0.00147	3.5	0.003
KHA_5O_4	0.1728	2.198	-0.03360	1.0	0.001
K_2HPO_4	0.0330	1.699	0.03090	1.0	0.002
$K_2Pt(CN)_4$	0.0881	3.164	0.02470	1.0	0.005
K_2SO_4	0.0666	1.039	---	0.7	0.002
Li_2SO_4	0.1817	1.694	-0.00753	3	0.002
$MgBr_2$	0.5769	2.337	0.00589	5	0.004

Table 2 (Continued)

Species	$4/3\,\beta_0$	$4/3\,\beta_1$	$2^{5/2}/3\,C^\phi$	Max m	σ
$MgCl_2$	0.4698	2.242	0.00979	4.5	0.003
$Mg(ClO_4)_2$	0.6615	2.678	0.01806	2	0.002
MgI_2	0.6536	2.4055	0.01496	5	0.003
$Mg(NO_3)_2$	0.4895	2.113	-0.03889	2	0.003
$MnCl_2$	0.4363	2.067	-0.03865	2.5	0.003
$(NH_4)_2SO_4$	0.0545	0.878	-0.00219	5.5	0.004
Na_2CO_3	0.2530	1.128	-0.09057	1.5	0.001
Na_2CrO_4	0.1250	1.826	-0.00407	2	0.002
$Na_2HA_5O_4$	0.0407	2.173	0.0034	1	0.001
Na_2HPO_4	-0.0777	1.954	0.0554	1	0.002
$Na_2S_2O_3$	0.0882	1.701	0.00705	3.5	0.002
Na_2SO_4	0.0261	1.484	0.01075	4	0.003
$NiCl_2$	0.4639	2.108	-0.00702	2.5	0.002
$Pb(ClO_4)_2$	0.4443	2.296	-0.01667	6	0.004
$Pb(NO_3)_2$	-0.0482	0.380	0.01005	2	0.002
Rb_2SO_4	0.0772	1.481	-0.00019	1.8	0.001
$SrBr_2$	0.4415	2.282	0.00231	2	0.001
$SrCl_2$	0.3810	2.223	-0.00246	4	0.003
$Sr(ClO_4)_2$	0.5692	2.089	-0.02472	2.5	0.003
SrI_2	0.5350	2.480	-0.00501	2	0.001
$Sr(NO_3)_2$	0.1795	1.840	-0.03757	2	0.002
UO_2Cl_2	0.5698	2.192	-0.06951	2	0.001
$UO_2(ClO_4)_2$	0.8151	2.859	0.04089	2.5	0.003
$UO_2(NO_3)_2$	0.6143	2.151	-0.05948	2	0.002
$ZnBr_2$	0.6213	2.179	-0.2035	1.6	0.007
$ZnCl_2$	0.3469	2.190	-0.1659	1.2	0.006
$Zn(ClO_4)_2$	0.6747	2.396	0.02134	2	0.003
ZnI_2	0.6428	2.594	-0.0269	0.8	0.002
$Zn(NO_3)_2$	0.4641	2.255	-0.02955	2	0.001

Table 3: 3-1 Electrolytes

Species	$3/2\ \beta_0$	$3/2\ \beta_1$	$3^{5/2}\ C^\phi$	Max m	σ
$AlCl_3$	1.0490	8.767	0.0071	1.6	0.005
$CeCl_3$	0.9072	8.400	-0.0746	1.8	0.007
$Co(en)_3Cl_3$	0.2603	3.563	-0.0916	1.0	0.003
$Co(en)_3(ClO_4)_3$	0.1619	5.395	---	0.6	0.007
$Co(en)_3(NO_3)_3$	0.1882	3.935	---	0.3	0.010
$Co(pn)_3(ClO_4)_3$	0.2022	3.976	---	0.3	0.003
$CrCl_3$	1.1046	7.883	-0.1172	1.2	0.005
$Cr(NO_3)_3$	1.0560	7.777	-0.1533	1.4	0.004
$DyCl_3$	0.9290	8.400	-0.0456	3.6	0.005
$Dy(ClO_4)_3$	1.2010	9.800	0.0142	2.0	0.006
$Er\ Cl_3$	0.9285	8.400	-0.0389	3.7	0.006
$Dr(ClO_4)_3$	1.2020	9.800	0.0144	1.8	0.004
$Er(NO_3)_3$	0.9380	7.700	-0.2260	1.5	0.006
$EuCl_3$	0.9115	8.400	-0.0547	3.6	0.006
$Ga(ClO_4)_3$	1.2381	9.794	0.0904	2.0	0.008
$GdCl_3$	0.9139	8.400	-0.0494	3.6	0.006
$Gd(ClO_4)_3$	1.1730	9.800	0.0140	2.0	0.007
$Gd(NO_3)_3$	0.7760	7.700	-0.1700	1.4	0.005
$HoCl_3$	0.9376	8.400	-0.0450	3.7	0.006
$Ho(ClO_4)_3$	1.1980	9.800	0.0132	2.0	0.004
$InCl_3$	-1.6800	-3.850	---	0.01	---
K_3AsO_4	0.7491	6.511	-0.3376	0.7	0.001
$K_3Co(CN)_6$	0.5603	5.815	-0.1603	1.4	0.008
$K_3Fe(CN)_6$	0.5035	7.121	-0.1176	1.4	0.003
K_3PO_4	0.5594	5.958	-0.2255	0.7	0.001

Table 4: 2-2 Electrolytes (b=1.2, α_1=1.4, α_2=12.)

Species	β_0	β_1	β_2	C^ϕ	range	σ
$BeSO_4$	0.3170	2.914	?	0.0062	.1-.4	0.004
$CaSO_4$	0.2000	2.650	-55.70	---	0.004-.011	0.003
$CdSO_4$	0.2053	2.617	-48.07	0.0114	.005-3.5	0.002
$CoSO_4$	0.2000	2.700	-30.70		.006-0.1	0.003
$CuSO_4$	0.2340	2.527	-48.33	0.0044	.005-1.4	0.003
$MgSO_4$	0.2210	3.343	-37.23	0.0250	.006-3.0	0.004
$MnSO_4$	0.2010	2.980	?	0.0182	0.1-4.0	0.003
$NiSO_4$	0.1702	2.907	-40.06	0.0366	.005-2.5	0.005
UO_2SO_4	0.3220	1.827	?	-0.0176	0.1-5.0	0.003
$ZnSO_4$	0.1949	2.883	-32.81	0.0290	.005-3.5	0.004

APPENDIX II

The Pauling Crystalline Radii*

Ion	radius (nm)	Ion	radius (nm)
Li^+	0.068	La^{+3}	0.112
Na^+	0.095	Pr^{+3}	0.115
K^+	0.134	Nd^{+3}	0.107
Cs^+	0.169	Sm^{+3}	0.106
Rb^+	0.148	In^{+3}	0.081
Ag^+	0.126	Tl^{+3}	0.095
Cs^+	0.166	B^{+3}	0.027
Cu^+	0.077	Cu^{+3}	0.054
Au^+	0.137	P^{+3}	0.044
NH_4^+	0.166	Eu^{+3}	0.106
OH_3^+	0.113	Gd^{+3}	0.100
Be^{++}	0.031	N^{+3}	0.016
Mg^{++}	0.072	Tb^{+3}	0.102
Ca^{++}	0.103	Dy^{+3}	0.101
Cr^{++}	0.073-0.080	Cr^{+5}	0.049
Sr^{++}	0.113	Lu^{+3}	0.095
Ba^{++}	0.135	Er^{+3}	0.098
Hg^{++}	0.110	Tm^{+3}	0.097
Mn^{++}	0.081	C^{+4}	0.016
Fe^{++}	0.061-0.078	Si^{+4}	0.040
Co^{++}	0.070	S^{--}	0.184
Ni^{++}	0.068	O^{--}	0.140
Zn^{++}	0.074	F^-	0.135
Cu^{++}	0.073	Cl^-	0.181
Cd^{++}	0.095	Br^-	0.194
Al^{+3}	0.049	I^-	0.222
Cr^{+3}	0.060	NO_3^-	0.206
Fe^{+3}	0.055-0.065	SO_4^{-2}	0.240
N^{+3}	0.016	P^{+3}	0.044
N^{+5}	0.013	P^{+5}	0.038

*References: (i) Y. Marcus, J. Solution Chem. 12, 271 (1983). (ii) H.S. Harned, B.B. Owen, "Physical chemistry of electrolyte solutions" (Reinhold Pub. New York, 1950), Table 5.1.6. (iii) L. Pauling,. *The Nature of the Chemical Bond* (3rd Ed.).(Cornell University Press, Ithaca, NY., 1960).

APPENDIX III

The Dielectric Constants of Selected Liquid Solvents

Solvent	Dielectric constant, D	T°C
Carbon tetrachloride	2.238	20
Tetranitromethane	2.52	25
Carbon dioxide	1.60	0
Carbon disulfide	2.641	20
Bromoform	4.39	20
Chloroform	4.806	20
Dibromomethane	7.77	10
Dichloromethane	9.08	20
Formic acid	58.0	16
Chloromethane	12.6	-20
Formamide	109.	20
Methane	1.70	-173
Methanol	32.63	25
Ethylene Glycol	37	25
Methylamine	11.4	-10
Dimethylamine	6.32	0
Trichloroethylene	3.4	16
Ethyleneoxlde	13	-1
Acetaldehyde	21.	10
Acetic acid	6.15	20
Acetamide	59.	83
Nitroethane	28.	30.
Ethanol	24.30	25
Methylether	5.02	25
Ammonia	16.9	25
Nitrogen	1.454	-203
Oxygen	1.507	-193
Hydrogen fluoride	17	-73
Hydrogen chloride	3.39	-50
Chlorine	2.10	-50
Carbon dioxide	1.60	20
Cyclohexane	2.023	20
Benzene	2.284	20
1-Propanol	20.1	25
Pyrrol	7.48	18
Furan	2.95	25
Hydrocyanic acid (CHN)	114	20
Aniline	6.89	20

Dielectric Constant of Water

From NSRDS-NBS 24
W. J. Hamer

T°C	Dielectric D	T°C	Dielectric, D
0	87.90	50	69.88
5	85.90	55	68.30
10	83.95	60	66.76
15	82.04	65	65.25
18	80.93	70	63.78
20	80.18	75	62.34
25	78.36	80	60.93
30	76.58	85	59.55
35	74.85	90	58.20
38	73.83	95	56.88
40	73.15	100	55.58
45	71.50		

*D= relative dielectric constant = $\varepsilon_m/\varepsilon_0$= permittivity of medium/permittivity of vacuum.

APPENDIX IV

Experimental* Mean Activity Coefficients of Aqueous NaCl Solution at 25°C

Molal	$\gamma\pm$	Molal	$\gamma\pm$
.001	.96511	.3	.70907
.002	.95189	.4	.69269
.003	.94220	.5	.68118
.004	.93431	.6	.67282
.005	.92756	.7	.66667
.006	.92162	.8	.66217
.007	.91628	.9	.65895
.008	.91142	1.0	.65677
.009	.90695	1.5	.65660
.01	.90280	2.0	.66818
.02	.87189	2.5	.68779
.03	.85094	3.0	.71392
.04	.83489	3.5	.74589
.05	.82182	4.0	.78337
.06	.81080	4.5	.82618
.07	.80126	5.0	.87427
.08	.79288	5.5	.92757
.09	.78539	6.0	.98604
.1	.77865		
.2	.73405		

*NBS Data.

Experimental* Mean Activity Coefficients of Aqueous KOH Solution at 25°C

Molal	$\gamma\pm$	Molal	$\gamma\pm$
.001	.96497	4.5	1.50101
.005	.92704	5.0	1.69717
.01	.90196	5.5	1.92293
.03	.84962	6.0	2.18227
.05	.82064	6.5	2.47962
.07	.80056	7.0	2.81994
.09	.78539	7.5	3.20874
.1	.77906	8.0	3.65204
.2	.74003	8.5	4.15644
.3	.72207	9.0	4.72904
.4	.71348	9.5	5.37745
.5	.71029	10.0	6.10974
.6	.71065	11.0	7.85990
.7	.71357	12.0	10.04903
.8	.71845	13.0	12.74508
.9	.72491	14.0	16.00624
1.0	.73268	15.0	19.87002
1.5	.78654	16.0	24.33974
2.0	.85984	17.0	29.36964
2.5	.95022	18.0	34.85075
3.0	1.05787	19.0	40.60036
3.5	1.18413	20.0	46.35881
4.0	1.33101		

*NBS Data.

APPENDIX V
Experimental Data* in Acid Gas Treating

Table V.1. Experimental Data of H_2S in 20 wt % MEA Solution

System Temperature (C)	System Total Pressure (kPa)	H_2S Partial Pressure (kPa)	Mole Ratio in Liquid (H_2S/MEA)
40.0	200.0	0.00164	0.00401
	201.0	0.00221	0.00427
	198.3	0.00290	0.00623
	197.5	0.0140	0.0274
	197.7	0.0221	0.0283
	195.7	0.0179	0.0322
	196.3	0.449	0.186
	783.6	777.	1.16
	1956.	1949.	1.40
	3845.	3838.	1.58 (LLE)
70.0	197.6	0.00520	0.00363
	197.6	0.00564	0.00367
	197.7	0.102	0.0281
	197.7	0.0991	0.0288
	198.0	2.27	0.196
	811.8	783.	1.03
	1819.	1789.	1.19
	3915.	3885.	1.46
100.0	266.6	0.0155	0.00390
	266.7	0.338	0.0280
	266.6	9.16	0.190
	783.2	688.	0.908
	1845.	1750.	1.06
	3714.	3619.	1.27
120.0	300.9	0.0288	0.00397
	300.7	0.752	0.0271
	302.3	18.5	0.180
	768.7	583.	0.745
	1845.	1659.	0.979
	3582.	3397.	1.16

*Data from D.B. Robinson Ltd. (GPA RR Reports)

Table V.2. Experimental Data of H_2S in 23.1 wt % MDEA Solution

System Temperature (C)	System Total Pressure (kPa)	H_2S Partial Pressure (kPa)	Mole Ratio in Liquid (H_2S/MDEA)
40.0	194.2	0.0033	0.00365
	197.3	0.0707	0.0282
	194.5	2.34	0.204
100.0	266.9	0.0391	0.00418
	265.5	0.846	0.0286
	263.9	21.5	0.183
120.0	300.4	0.0555	0.00341
	300.0	1.68	0.0252
	300.8	35.1	0.154

Table V.3. H_2S in Solutions of 35 wt % MDEA + 5 wt % DEA

System Temperature (C)	System Total Pressure (kPa)	H_2S Partial Pressure (kPa)	Mole Ratio in Liquid (H_2S/Amines)
40.0	196.5	0.00476	0.00331
	197.7	0.160	0.0287
	195.7	3.76	0.192
	858.9	852.	1.14
	1810.	1803.	1.40
	2865.	2859.	1.66
70.0	197.1	0.0133	0.00338
	196.6	0.703	0.0298
	197.2	12.7	0.182
	748.9	721.	0.963
	1845.	1817.	1.20
	3707.	3679.	1.51
100.0	266.9	0.0483	0.00307
	266.1	2.09	0.0274
	267.4	18.94	0.109
	265.6	30.42	0.154
120.0	300.1	0.0938	0.00289
	300.5	3.98	0.0255
	300.2	33.4	0.0987
	300.3	48.3	0.128
	755.3	575.	0.477
	1955.	1775.	0.854
	3624.	3444.	1.08

Table V.4. Experimental Data of CO_2 in 20 wt % MEA Solution

System Temperature (C)	System Total Pressure (kPa)	CO_2 Partial Pressure (kPa)	Mole Ratio in Liquid (CO_2/MEA)
40.0	233.0	0.00330	0.122
	195.9	0.255	0.427
	161.3	154.	0.712
	783.9	777.	0.875
	2163.	2156.	1.00
	6300.	6293.	1.18
70.0	266.7	0.0713	0.125
	265.8	5.19	0.430
	173.7	144.	0.612
	769.6	740.	0.738
	2162.	2133.	0.858
	6306.	6274.	1.02
100.0	301.2	1.22	0.133
	438.6	36.2	0.398
	246.4	152.	0.522
	887.4	793.	0.646
	2253.	2158.	0.744
	6383.	6288.	0.893
120.0	540.6	4.25	0.119
	679.9	112.	0.369
	312.9	127.	0.419
	878.5	693.	0.554
	2258.	2072.	0.661
	6388.	6203.	0.800

Table V.5. Experimental Data of CO_2 in 30 wt % DEA Solution

System Temperature (C)	System Total Pressure (kpa)	CO_2 Partial Pressure (kpa)	Mole Ratio in Liquid (CO_2/DEA)
40.0	197.6	0.00169	0.0172
	197.4	0.279	0.221
	159.4	153.	0.746
	763.7	757.	0.940
	2177.	2170.	1.06
	6286.	6280.	1.22
70.0	347.6	0.00170	0.00265
	215.6	0.0411	0.0205
	267.5	4.34	0.218
	162.3	133.	0.590
	783.1	754.	0.789
	2169.	2140.	0.930
	6382.	6353.	1.09
100.0	490.0	0.0238	0.00409
	369.4	0.398	0.0190
	424.5	0.467	0.0211
	397.0	35.4	0.229
	231.2	136.	0.378
	865.5	770.	0.598
	2245.	2150.	0.755
	6368.	6273.	0.935
120.0	531.8	0.0837	0.00302
	611.2	1.458	0.0166
	645.5	84.4	0.174
	231.4	45.3	0.140
	886.4	700.	0.440
	2252.	2066.	0.591
	6382.	6195.	0.809

Table V.6. Experimental Data of CO_2 in 23 wt % MDEA Solution

System Temperature (C)	System Total Pressure (kPa)	CO_2 Partial Pressure (kPa)	Mole Ratio in Liquid (CO_2/MDEA)
40.0	274.6	0.00200	0.00334
	232.1	0.0414	0.0188
	196.3	2.71	0.230
	179.2	172.	0.942
	806.6	799.	1.05
	994.0	987.	1.08
	2168.	2161.	1.17
	2273.	2266.	1.18
	5273.	5265.	1.34
70.0	238.6	0.0656	0.00707
	265.7	0.282	0.0195
	265.3	17.9	0.205
	197.4	167.	0.679
	845.5	815.	0.927
	2328.	2297.	1.09
	5144.	5114.	1.22
100.0	507.4	0.113	0.00368
	440.1	1.42	0.0170
	241.4	143.	0.334
	927.5	829.	0.642
	2368.	2270.	0.875
	5251.	5152.	1.07
120.0	500.4	1.090	0.00705
	714.7	2.96	0.0130
	304.0	111.	0.185
	404.2	211.	0.241
	967.6	774.	0.432
	2476.	2283.	0.693
	5234.	5040.	0.903

BIBLIOGRAPHY

[Alphabetical listing with annotations]

[1] Allnatt, A.R., Molec. Phys. 8, 533 (1964).

[2] Astarita, G., D.W. Savage, A. Bisio, *"Gas treating with chemical solvents"* (Wiley, New York 1983).

[3] Attard, P., *"Electrolytes and the electric double layer"*, Adv. Chem. Phys. 92, 1 (1996).

[4] Austgen, D.M. and G.T. Rochelle, Ind. Eng. Chem. Res. 28, 1060 (1989). [Enhancement factor]

[5] Bacquet, R., P.J. Rossky, J.Chem.Phys. 79, 1419 (1983). [soft RPM]

[6] Barrat, J.L., J.P. Hansen, G. Pastore, Mol. PHys. 63, 747 (1988) [Triplet direct correlations $C^{(3)}$]

[7] Bell, G.M., S. Levine, *"Chemical Physics of Ionic Solutions"* ed. B.E Conway, R.G. Barradas (Wiley, NY, 1966). [Modified Poisson-Boltzmann eqautions]

[8] Blum, L., *"Primitive electrolytes in the mean spherical approximation"*, in *"Theoretical chemistry: Advances and perspectives"*, ed. H. Eyring and D. Henderson (Academic Press, New York, 1980).

[9] Blum, L., *"Structure of the Electric Double Layer"* Adv. Chem. Phys. 78, 171 (1990).

[10] Blum, L. and J.S. Hoye, J. Phys. Chem. 81, 1311 (1977).

[11] Born, M., Z. Phys. 1, 45 (1920). N. Bjerrum, *"Selected Papers"*, (Munskgaard, Copenhagen, 1949).

[12] Boublik, T. Molec. Phys. 27, 1415 (1974). J. Chem. Phys. 63, 4084 (1975). [Equation of state for convex bodies]

[13] M. Broccio, D. Costa, Y. Liu, S.H. Chen, J. Chem. Phys. 124, 084501 (2006). [Structure factors of proteins in salt solutions]

[14] Broul, M., K. Hlavaty, J. Linek, Collect. Czech. Chem. Commun. 34, 3428 (1969). [Experiment data of methanol-water-LiCI at 60°C]

[15] Cardoso, M. and J.P. O'Connell, Fluid Phase Equil. 33, 315 (1987). [McMillan-Mayer and Lewis-Randall scales]

[16] Carley, D.D., J. Chem. Phys.46, 3783 (1967) [HNC solution for electrolytes]

[17] Carnie, S. L., G. M. Torrie, *"The statistical mechanics of electrical double layer"*, Adv. Chem. Phys. 58, 141 (1984). [Review]

[18] Chakravarty, T., *"Solubility calculations for acid gases in amine blends."* Ph.D. Thesis (Clarkson University, Potsdam NY, 1985).

[19] Chapman, W. G., M. Llano-Restrpo, J. Chem. Phys. 100, 5139 (1994). [Lennard-Jones renormalized bridge function]

[20] Chapman, D.L., Phil. Mag. 25, 475 (1913).

[21] Corson, D.R., P. Lorrain. *"Introduction to electromagnetic fields and waves"* (W.H. Freeman and Company, San Francisco, 1962). 1st ed.

[22] Curtis, R.A., J.M. Prausnitz, H.W. Blanch, Biotech. & Bioeng. 57, 11 (1998). [Aquesou protein-electrolyte solutions.]. Curtis, R.A, L. Lue, Chem. Eng. Sci. 61, 907 (2006). [Biosepration of proteins in salt solution]

[23] Debye, P., E. Hückel, Z. Physik 24, 185 and 305 (1923).

[24] Derjaguin, B.V., L.D. Landau, Acta Physicochim. 14, 633 (1941). [DLVO theory]

[25] Deshmukh, R.D., A.E. Mather, Chem. Eng. Sci. 36, 355 (1981). [Acid gas equations]

[26] Duh, D.M., A.D.J. Haymet, J. Chem. Phys. 97, 7716 (1992). [Integral equations for electrolytes]

[27] Duh, D.M., D. Henderson, J. Chem. Phys. 104, 6742 (1996). [$\Delta\gamma$ for B(r)]

[28] Dumetz, A. C., A. M. Snellinger-O'Brien, E. W. Kaler and A. M. Lenhoff, Protein Sci. 16, 1867 (2007). [Protein crystallization]

[29] Fredenslund, A., J. Gmehling, P. Rassmussen, *"Vapor-liquid equilibrium using UNIFAC"* (Elsevier, Amsterdam 1977). [UNIFAC]

[30] Friedman, H.L., *"A course in statistical mechanics"* (Prentice-Hall, Englewood Cliffs, New Jersey, 1985); and *"Ionic solution theory based on cluster expansion methods"* (Interscience, New York, 1962). [Books]. H.L. Friedman, J. Chem. Phys. 32, 1351 (1960); and J. Solution Chem. 1, 387, 413, 419 (1972) [Scale conversion]. W.D.T. Dale, H.L. Friedman, J. Chem. Phys. 68, 3391 (1978).

[31] Furter, W.F., Ph.D. Thesis, University of Toronto (1958).

[32] Furter, W.F. and R. A. Cook, Salt effect in distillation: a literature review. Int. J. Heat Transfer, 10, 23 (1967).

[33] Furter, W.F., Extractive distillation by salt effect. ACS Symp. Ser. No. 155: 35-45 (1972).

[34] Furter, W.F., *"Thermodynamic behavior of electrolytes in mixed solvents"*, Advances in Chemistry Series 155 (American Chemical Society, Washington D.C., 1976).

[35] Furter, W.F., *"Thermodynamic behavior of electrolytes in mixed solvents II"*, Advances in Chemistry Series 177 (American Chemical Society, Washington D.C., 1979).

[36] Gering, K.L., *"A molecular approach to electrolyte solutions: Predicting phase behavior and thermodynamic properties of single and binary-solvent systems"* Doctoral Dissertation., (University of Oklahoma, 1989).

[37] Gering, K.L., L.L. Lee, L.H. Landis, Fluid Phase Equil. 48, 111 (1989). [Phase equilibria of binary solvents with MSA]

[38] Gering, K.L., L.L. Lee, Fluid Phase Equil. 53, 199 (1989). [Cosolvent electrolytes with MSA]

[39] Gouy, L., *"Sur la constitution de la charge électrique à la surface d'un électrolyte"* J. Physique (Paris) 9, 457 (1910).

[40] Grahame, D.C., Chem. Rev. 41, 441 (1947).

[41] Hafskjold, B., G. Stell, in *"The liquid state of matter: Fluids, Simple and Complex"*, ed. E.W. Montroll, J.L. Lebowitz (North-Holland, New York, 1982).

[42] Hansen, J.P., Zerah, G. Phys. Lett. 108A, 277 (1985). G. Zerah, J.P. Hansen, J. Chem. Phys. 84, 2336 (1986) [Bridge functions with renormalized potentials]

[43] Harned , H.S., B.B. Owen, *"Physical chemistry of electrolyte solutions"* (Reinhold Pub. New York, 1950)

[44] Harvey, A.H., J.M. Prausnitz, AIChE J., 335: 635 (1989). [High-pressure aqueous systems containing gases and salts]. A.H. Harvey, J.M. Prausnitz, J. Solution Chem. 16, 857 (1987). [Dielectric constants in mixed solvents]

[45] Helmholtz, H., Prog. Ann. 89, 211 (1853). [Helmholtz layer in EDL.]

[46] Henderson, D., F.F. Abraham, J.A. Berker, Mol.Phys. 31, 1291(1976). [Ornstein-Zernike equation for nonuniform fluids]

[47] Høye , J.S., L. Blum, J. Stat. Phys. 16, 399 (1977).

[48] Israelachvili, J.N. *"Intermolecular and surface forces."* (Academic Press, New York 1985)] [Electric double layers and DLVO forces.]

[49] Johnson, A.I., Furter, W.F. Can. J. Chem. Eng. 38,78 (1960). Ibid. 43, 356 (1965). [Salt effects in mixed solvent electrolytes.]

[50] Joly, L., C. Ybert, E. Triaze, L. Bocquet, Phys.Rev. Lett. 93,257805 (2004) [Molecular dynamics simulation of ζ–potential.]

[51] Kent, R.L. and B. Eisenberg, Hydrocarbon Processing 55, 87 (1976).

[52] Kirkwood, J.G. and Buff, F.P., J. Chem. Phys., 19: 774.(1951). [Solution theory]. J.G. Kirkwood, J. Chem. Phys. 7, 911 (1939). [Dielectric constant]

[53] Kjellander, R., S. Marcelja, Chem. Phys. Lett. 112, 49 (1984); J. Chem. Phys. 82, 2122 (1985). Chem. Phys. Lett. 127, 402 (1986). [Kirkwood equation for EDL]

[54] Kohl, A.L. and F.C. Riesenfeld, *"Gas Purification"* (McGraw-Hill, New York, 1960).

[55] Labik, S., A. Malijevski, P. Vonka, Molec. Phys. 56, 709 (1985). [Newton-Raphson method of solution of integral equations with Fourier series.]

[56] Landis, L.H., *"Mixed-salt electrolyte solutions: accurate correlations for osmotic coefficients based on molecular distribution functions."* Ph.D. Dissertation (University of Oklahoma, 1985).

[57] Lee, L.-J. *"A vapor-liquid equilibrium model for natural gas sweetening process"* Ph.D. Dissertation (The University of Oklahoma, 1996).

[58] Lee. L.L., J. Chem. Phys. 78, 5270 (1983). [MSA coupling parameters.]

[59] Lee, L.L., *"Molecular thermodynamics of nonideal fluids"* (Butterworths, Boston, 1988).

[60] Lee. L.L., J. Chem. Phys. 97, 8608 (1992). [Direct chemical potential formula]. *Ibid.* 103, 4221 (1995). [Zero-separation theorems]

[61] Lee, L. L., J. Chem. Phys. 104, 8058 (1996). [Renormalized $\Delta\gamma$ in the closure.]

[62] Lee, L.L., L.-J. Lee, D. Ghonasgi, M. Llano-Restrepo, W.G. Chapman, K.P. Shukla, E. Lomba, Fluid Phase Equil. 116, 185 (1996). [Mixed-solvent phase equilibria]

[63] Lee, L.L., Fluid Phase Equil. 131, 67 (1997). [Setchenov salting out].

[64] Lee, L.L., Fluid Phase Equil. 158, 401 (1999). [Absorption refrigeration.]

[65] Lee, L.L., J. Molec. Liquids 87, 129 (2000). [Scale conversion McMillan-Mayer to Lewis-Randall]

[66] Lobo, V.M.M., Quaresma, J.L., *"Electrolyte solutions: Literature data on thermodynamic and transport properties"* (Universidade de Coimbra, Portugal, 1981).

[67] Loeb, A.L., J. Colloid. Sci. 6, 75 (1951).

[68] Long, F.A. and McDevit, W.F.,. Chem. Rev., 51: 119 (1952).[Activity coefficients of neutral solutes in salt solutions]

[69] Lonetti, B., E. Fratini, S.H. Chen, P. Baglioni, PCCP 6, 1388 (2004). [SANS Cytochrome C]. Y. Liu, E. Fratini, P. Baglioni, W.R. Chen, S.H. Chen, Phys. Rev. Lett. 95, 118102 (2005). [Protein clusters]

[70] Lovett, R., C.Y. Mou, F.P. Buff, J. Chem. Phys. 65, 2377 (1976).

[71] Lyklema, J. *"Foundations of colloidal and interface science"* (Academic Press, San Diego 1995), Volume 2.

[72] Marcus, Y., *"Ion Solvation"*. (Wiley-Interscience, Chichester, 1985).

[73] Martynov, G. A., G. N. Sarkisov, Mol. Phys.49, 1495 (1983). [Thermal function in closure]

[74] Matteoli, E., L. Lepori, J. Chem. Phys., 80: 2856. (1984). [Solute-solute interactions in water] [Analysis through the Kirkwood-Buff integrals for 14 organic solutes]

[75] Mock, B., L.B. Evans, C.C. Chen,. *AIChE Journal*, 32, 1655 (1986). [Mixed Solvent Electrolyte Systems.]

[76] Morita, T., Progr. Theor. Phys., 20, 920 (1958). [Theory of classical fluids: Hypernetted chain approximation]

[77] Neilson, G.W. and Enderby, J.E., Ann. Rep. Prog. Chem. Sect. C, 76: 185 (1979).

[78] O'Connell, J.P., in E. Matteoli and G.A. Mansoori (Eds.), *"Fluctuation Theory of Mixtures"*. (Taylor and Francis, New York, 1990) pp. 45-67.

[79] Outhwaite, C.W., Chem. Phys. Lett., 7, 639 (1970).

[80] Outhwaite, C.W., L.B. Bhuiyan, S. Levine, J. Chem. Soc. Faraday Trans. II 76, 1388 (1980).

[81] Pawlikowski, E.M. and Prausnitz, J.M., Ind. Eng. Chem. Fundam., 22, 86 (1983). [Salting out]

[82] Perram, J.W., E.R. Smith, Chem. Phys. Lett. 39, 328 (1976). [Non-uniform OZ equation]

[83] Pitzer, K.S., J. Phys. Chem. 77: 268 (1973). Pitzer, K.S., G. Mayorga, J. Phys. Chem. 77, 2300 (1973). [Pitzer's virial electrolyte formulas]

[84] Pitzer , K.S., L.F. Silvester, J. Phys. Chem. 81, 1822 (1977) [Heat of Solution]

[85] Plischke, M., D. Henderson, J. Phys. Chem. 88, 2712 (1988). [BGY equation in EDL]

[86] Rasaiah, J.C, H.L. Friedman, J. Chem. Phys. 48, 2742 (1968). J.C. Rasaiah, H.L. Friedman, J. Chem. Phys. 50, 3965 (1969). J.C. Rasaiah, J. Chem Phys. 56, 3071 (1972). J.C. Rasaiah, J. Solution Chem. 2, 301 (1973).

[87] Reed, T.M., K.E. Gubbins, *"Applied Statistical Mechanics"*, (McGraw-Hill, New York 1973).

[88] Robinson, R.A., Stokes, R.H., *"Electrolyte Solutions"* (Butterworths, London, 1959).

[89] Rogers, F.A, D.A. Young, Phys. Rev. A30, 999 (1984). [New Closure]

[90] Y. Rosenfeld, N.W. Ashcroft, Phys. Rev. A28,1208 (1979)

[91] Rossky, P.J., J.B. Dudowicz, B.L. Tembe, and H.L. Friedman, J. Chem. Phys. 73, 3372 (1980). P.J. Rossky, H.L. Friedman, J. Chem. Phys. 72, 5694 (1980). P.J. Rossky, J. Chem. Phys. 73, 2457 (1980) [renormalization]

[92] Sciortino, F., S. Mossa, E. Zaccarelli, P. Tatarglia, Phys. Rev. Lett. 93, 055701 (2004). [Protein clusters]

[93] Setchenov, J. Z. Phys. Chem. 4. 117 (1889) [Salting out]

[94] Simonin, J.-P. Chem. Soc. Fraraday Trans. 92, 3519 (1996). [Scale conversion]. J. P. Simonin, L. Blum, P. Turq, J. Phys. Chem. 100, 7704 (1996). J. P. Simonin, O. Bernard, L. Blum, J. Phys. Chem. B 102, 4411 (1998). J. P. Simonin, O. Bernard, L. Blum, J. Phys. Chem. B 103, , 699 (1999). T. Vilarino, O. Bernard, J.-P. Simonin, J. Phys. Chem. B 108, 5763 (2004). [MSA for salt solutions]

[95] Smith, W.R., D. Henderson, Y. Tago, J. Chem. Phys. 67, 5308 (1977). W.R. Smith, D. Henderson, J. Chem. Phys. 69, 319 (1978)

[96] Spaarnay, J., *"The Electric Double Layer"* (Pergamon, Oxford, 1972).

[97] Stern, O. Z. Elektrochem, 30, 508 (1924).

[98] Stillinger, F.H., R. Lovett, J. Chem. Phys. 48, 3858 and 3869 (1968) [2nd moment condition]

[99] Stradner, A., H. Sedgwick, F. Cardinaux, W.C.K. Poon, S.U. Egelhaaf, P. Schurtenberger, Nature, 432, 492 (2004). [SANS & SAXS of proteins]

[100] Tasseven, C., L.E. Gonzalez, M. Silbert, O. Alcaraz, J. Trullas, J. Chem. Phys. 115, 4676 (2001)] [Molten salts]

[101] Valleau, J.P., L.K. Cohen, D.N. Card, J. Chem. Phys. 72, 5942 (1980). D.N. Card, J.P. Valleau, J. Chem. Phys. 52, 6232 (1970).

[102] Van Leeuwen, J.M., J. Groeneveld, J. de Boer, Physica 25, 792 (1959). Verlet, L., Nuovo Cimento 18, 77 (1960). E. Meeron, J. Math. Phys. 1, 192 (1960). T. Morita, K. Hiroike, Progr. Theor. Phys. (Kyoto) 23, 1003 (1960). [HNC integral equation].

[103] Verwey, E.J.W., J.T.G.Overbeek, *"Theory of the stability of lyophobic colloids"* (Elsevier, Amsterdam, 1948).

[104] Waisman, E., J.L. Lebowitz, J. Chem. Phys. 52, 4037 (1970). *Ibid.* 56, 3086 (1972). [MSA solutions]

[105] Walters, H., D.E. Brooks, D. Fisher (eds.) *"Partitioning in aqueous two-phase systems"* (Academic Press, London, 1985).

[106] Wang, P., A. Anderko, Fluid Phase Equil. 186, 103 (2001).

[107] Weast, R. ed. *"Handbook of Chemistry and Physics"* (CRC Press, Cleveland, Ohio, 1974).

[108] Weiland, R.H., T. Chakravarty, and A.E. Mather, Ind. Eng. Chem. Res. 32, 1419 (1993).

[109] Weingaertner, D.E., Lynn, S., Hanson, D.N., Ind. Eng. Chem. Res. 30, 490 (1991). [Extractive crystallization]

[110] Wertheim, M.S., J. Chem. Phys. 65,2377 (1976). [WLMB equation]

[111] Wilczek-Vera, G., E. Rodil, J.H. Vera, Fluid Phase Equil. 244, 33 (2006). [Single ion activities]

[112] Wu, S.W. and S.I. Sandler, Ind. Eng. Chem. Res. 30, 881 and 889 (1991).

[113] Wu, R.S. and L. L. Lee, Fluid Phase Equil. 78, 1 (1992). [Mixed-solvent electrolytes with MSA.]

[114] Wu, R.S.,L.L. Lee, Fluid Phase Equil. 131, 67 (1997). [Mixed solvents]

[115] Zwanzig, R., Triezenberg, D.G., Phys. Rev. Lett. 28, 1183 (1972). [Non-uniform integral equations]

INDEX